Introduction to Biogeography

Introduction to Biogeography

BRIAN SEDDON

39,186

 Duckworth

First published in 1971 by
Gerald Duckworth and Co. Ltd.
3 Henrietta Street, London, WC2.

Cloth ISBN 0 7156 0586 0

Paper ISBN 0 7156 0587 9

IBM Typesetting by Specialised Offset Services, Liverpool.
Printed by Unwin Brothers Limited, Old Woking, Surrey.

PREFACE

Why Biogeography? There are at least two good reasons for believing that the subject should be better known, reasons that become more compelling with every year that passes.

For many kinds of plants and animals the threat of extinction grows as human demands increase on the limited resources of this planet. Man's power to change his living conditions is far reaching and already there is no part of the earth which is safe from the influences of his technology. We know now that extinction looms not merely in the form of a hunter with a gun but in more sinister guise, as unseen pesticides reaching the seas and exhaust gas poisons settling in the polar ice.

Conservation of threatened species used to be a plea but has become a necessity. Man himself cannot survive unless he ensures the continued existence of plant life and animal populations in all their variety. The message of the last two decades has been to point out the danger of monoculture and our dependence on diversity. Not so long ago the plants of field margins were weeds and the insects that lived there were pests but today their role in keeping a natural balance is more clearly seen. The unique qualities of each species are irreplaceable assets. Geographical uniqueness, as well as being important in its own right, is a key to understanding what role each plant and animal may be capable of fulfilling in peaceful co-existence with man. We are at last prepared to recognise that domestic cattle may not be the ideal meat-producing animals for East African plains but that native eland, springbok and zebra probably are. Every species has its place on this earth — its natural territory. This is where it has evolved, where it has survived changing fortunes in the slow procession of earth history and where its future potential may be assessed. That is why I consider that a discussion of species is the only credible way to begin in biogeography: it lays a foundation on which all else may be built.

This book is an attempt to explain in ordinary language the forces that have shaped the territories of living things and the problems that are posed by their geographical distribution. Much work has been published in the last twenty years in many branches of biology, ecology and geology which contributes to a new and more stimulating image for biogeography. To envelope all of it would be an impossible task but if the main themes emerge clearly from the selected work described a more limited task will have been achieved.

ACKNOWLEDGMENTS

I am indebted to the many friends and colleagues whose work is featured in these pages and whose views have influenced my thinking. Needless to say, I bear responsibility for the accuracy of statements attributed to others and for interpretations of them.

It is a particular pleasure to thank all those concerned in producing the maps which are so important a feature of the book, namely Mrs Mary Petts, Janet Brittlebank, Hugh Stewart-Killick, Mr I. Maclean and staff of the Reading University photographic service. Thanks are also due to Mrs Linda Todd for typing, Mrs Danuta Ganguli for translation and Miss Betzy Dinesen for editorial assistance. Acknowledgment for reproducing or adapting illustrations is due to the authors and publishers mentioned in the captions and bibliographies.

The text has benefitted from the reactions of student classes at Reading over several years and from discussion with practising teachers in schools. I hope it may now satisfy a wider audience in colleges and universities as well as justify its title for the general reader.

CONTENTS

CHAPTER I

Mapping distributions

Introduction

There can be no better way of introducing the subject of biogeography than to describe the sources of information that provide its basic material and to illustrate the methods of presenting these data in the form of maps. The mapping of distributions is an essential first step in biogeographic study because the aim is to relate the living organisms of the earth to the areas they occupy (and have occupied) and to try to understand this relationship. It is a bridge between the biological and the geographical sciences, gaining its unique character from the approach it adopts — fundamentally an areal one — to the history, and to the present and future roles of biological species on earth. It attempts to document and to compare over distances, within regions and between regions at all scales, the occurrence of plants and animals as individual species and as communities in relation to one another and to their environment. Material that cannot be represented in map form is not biogeography, but this is not to say that mapping is the only technique employed. Statistical methods of correlation, computerized arrangement into series (ordination) and into categories (classification) can all be used to examine the raw data where the quantity and quality of records make such treatment appropriate. In the final analysis, however, it is the concern with the spatial organisation of living things that distinguishes biogeography, and therefore only when information is expressed in relation to territorial occurrence does it become demonstrably biogeographical.

Sources of information

Apart from original survey in the field, the principal sources of information are institutes in which records are kept of flora and fauna in the form of specimens and documents. The primary record is the actual specimen, fully annotated at the time it was collected, for this is the only case in which the identity of the plant or creature can be checked subsequently if there is doubt about its accuracy. Museums, both national and local, are the storehouses of this information. Eventually, as material accumulates, much of it is published first in the form of regional and national catalogues. In their most basic form these are checklists of the flora or fauna of a particular country, e.g. *List of British Vascular Plants* (J. E. Dandy, 1958), in which the names of all the species known in that territory are listed in systematic order. In more elaborated form publication includes descriptions of every plant or animal listed, some account of their respective habitats and more or less detailed references to the areas or localities where each is found. Reference works of this kind usually deal with one major category of plants or animals, such as mosses, flowering plants, beetles, birds, mammals. In the case of botanical groups the publication is generally called a "flora", often a "handbook" in zoological groups. Additionally records and reports appear in the journals of national botanical and zoological societies and for smaller areas in the reports of local natural history societies. Accounts including fuller description of associated environments are published in the ecological and geographical journals.

Ultimately, when records are available in sufficient quantity for a wide area, it becomes possible to produce maps, and there are now available some excellent biogeographic atlases, at least for several countries and major continental regions, e.g. the British Isles, central Europe, Scandinavia, Alaska, Japan. In the search for information the biogeographer must be prepared to read the literature of botany, zoology, and ecology to obtain raw material if it does not already exist in map form. Even when distribution maps are available he will rely upon the ecological, geological and climatological literature of his area when he comes to examine the factors influencing those distributions.

There is great variation in the quality and quantity of information available from different parts of the world and one must be conscious of such limitations in the original data where they exist. The best documentation of wildlife is available in some of the scientifically advanced countries, but even among these a complete survey of natural resources, of which the variety of plant and animal life is an important aspect, may be still in progress, as in Australia under C.S.I.R.O. Countries that have a long tradition of interest in natural history possess a most comprehensive record of flora and fauna. On the other hand, those countries whose literate population is small in relation to their area possess least information and what is available may be the product of work done by visiting scientists of other nationalities, as in Turkey. The extreme situation in which all knowledge of the extent, and indeed of the existence, of many plant and animal species is fragmentary is found in territories that even today are unexplored. Topographic survey must precede all other kinds of exploration but since aerial photography can now provide the basis for the topographic map, biogeographic survey can be an important part even of reconnaissance ground expeditions.

Methods of survey

The range of survey methods that provide the basis of present knowledge can be illustrated by describing two extreme examples of the situation in an undeveloped and in an advanced country. New Guinea has been the subject of a number of special scientific expeditions, notably the six Archbold expeditions from the American Museum of Natural History between 1933 and 1959, the Dutch expedition sponsored by the Netherlands Geographical Society (1959) and the British Museum (Natural History) expedition of 1964-5 and 1969-70. When exploratory journeys are so few that they can be referred to individually it is obvious that accurate records are available only for the routes that these expeditions have covered. New Guinea could contain the Appalachians and its length would extend from New York to New Orleans. It is both mountainous and heavily forested presenting some of the most difficult terrain in the world to traverse, and consequently vast tracts of territory between these routes remain scientifically unexplored and unrecorded. In terms of distribution maps, therefore, known locations are few and are unevenly dispersed, so that the resulting map for any particular organism will be skeletal at best and leaves much to be inferred. Notice, incidentally, that information of some biogeographic value exists at two levels, one more superficial than the other. It is not too difficult to answer the question "Is there evergreen forest in the Star Mountains (West Irian)?" Nor would it be too difficult to state what elevation the forest reaches. Quite superficial observation by a non-specialist expedition can yield such information and, indeed, the development of air-photo interpretation obviates the need for ground expeditions to determine facts of this kind. However, it is quite another matter to answer questions such as "Are there beech forests (or tree-ferns, or screw-pines, or birds-of-paradise) in the Star Mountains?" At this level of enquiry, the specific level, there can be no alternative to specialist survey on the ground. Naturally, when definite answers can be given for at least some areas these tend to be regarded as samples representative of wider areas in which terrain and other conditions are believed to be similar.

It is possible to distinguish two stages in the evolution of surveys of flora and fauna in advanced countries with a long history of civilization. A tradition of amateur interest in natural history originated among small groups of professional men, such as apothecaries and physicians, during the eighteenth century and flourished in the later part of the nineteenth and early twentieth centuries. The individual usually concentrated on collecting or recording within an area that was accessible from his place of residence and practice, for many of these naturalists were clergymen or country doctors. Often the limits of a parish or a county were chosen to define the boundary of the collecting area. In any case these administrative divisions were used in reporting discoveries or specimens even if more precise location was not cited. With this kind of information as his source material, H. C. Watson first attempted a geographical assessment of the British flora by compiling an inventory for the whole country in which the presence or absence of every plant species was noted within eighteen provinces devised by himself. Thirty years later, in 1873, he published a revised catalogue in which distributions were described by reference to much smaller units which illustrated more subtle differences in the geographical occurrence of various species. He named his units "vice-counties", of which there were 112 (not including Ireland), and they were defined by the boundaries of the administrative counties except for the largest of these, e.g. Yorkshire, Devon, Norfolk, which he subdivided into units approximating in area to that of most smaller counties. Once such a system had been demonstrated it was adopted as a basis for further observations, records being added to Watson's register for vice-counties until the early 1950s.

A new basis for survey became available in 1954 when mechanised sorting of punched (i.e. perforated) index cards made it practicable to handle vastly greater amounts of information. This in turn made it possible to reduce the size of the unit area for which individual records were required, so that much greater detail in survey was achieved. However, the new survey that was launched in Britain and Ireland and carried through to completion within five years also depended on the fact that there were a large number of experienced amateur botanists who undertook the necessary field work voluntarily. The ten-kilometre square of the national grid formed the unit for recording observations, i.e. for most species a sighting had only to be referred to the appropriate square (10 x 10km) and these quite small areas of country were convenient units for conducting the search. Each square could be assigned to a particular observer and in this way, supplemented by group efforts to deal with more remote areas, 3,500 squares covering the whole country were examined, each yielding records for over 400 species on average. Compared with earlier methods this imposes more systematic coverage of the territory and results in the more regular spacing of observations. It thus improves recording in remote and under-populated areas. It also tends to counterbalance the over-intensive recording of accessible and attractive areas (which are often well-known in any case) so that the results as a whole are generally more representative. Thus, where a sufficient labour-force is available, a survey organised on a gridded division of the area produces the best return for effort and is less liable to artificial bias than surveys based upon divisions of irregular shape and size.

Having dealt with means of acquiring data and methods of conducting field survey, let us now consider methods of presenting this information in map form. The two things are distinct and separable stages in a process, and it is not necessary and sometimes not desirable to employ the same units to represent distribution on a map as those used for the organisation of the survey.

Taxus baccata L.

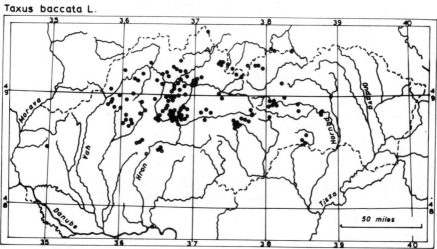

1.1   Distribution of European Yew in Slovakia (redrawn from *Flora Slovenska*). An example of dot map presentation at medium scale.

Methods of mapping distributions

A dot map has marked upon it the position of every point or locality where the plant or animal in question has been observed. The points are positioned with as much precision as the scale allows, and of course dots can be used to depict distributions on maps at any scale desired. Examples throughout this book illustrate its use in areas of very different size: first an English county (Dorset) (Fig. 3.1), second a major province (Slovakia) (Fig. 1.1) and third at world scale (Fig. 6.2). If we compare these, we see that both the number of records available and the size of the individual dot relative to the scale of the map influence the quality of the information conveyed. To be perfect, a dot map should include every individual point at which the species occurs, but in fact it can never do this (unless the species is so restricted in its total range that a few localities provide the only foothold for its existence). A dot map indicates that where a dot is placed the species is present, but by implication it suggests that where there is no dot the species is absent! Thus a map of this kind is excellent when survey is thorough and information complete because all positive records are precisely located and variations in frequency within the area are apparent (Fig. 3.1). But, whenever survey has not been complete or records from reliable sources are meagre, then the blank areas on a dot map may be misleading. For this reason dot maps have been described as honest but inclined to understatement!

Instead of attempting to show on a map every particular place at which the species occurs, it is a common practice to indicate its presence or absence within each administrative sub-division of the area. The use of British counties is an obvious example, for in the present century up to 1955 this unit was used in the preparation of maps as well as for the acquisition of records. Stapf (1916) was among the first to present maps in this form. The effect of using units larger than the individual locations of records is to generalize the information, which may be a reasonable course to adopt where a distribution is known to be under-recorded. Administrative units have been used for their convenience and because of their historical permanence. Their boundaries are already fixed and well documented so that description of distribution by counties will be widely understood and the area referred to readily recognised. The historical value of administrative units can scarcely be equalled by any other system, for old reports are often incomplete and localities may not be precisely stated, but usually they are sufficient to allow the record to be placed within the appropriate county or parish (cf. map of Wisconsin in Fig. 8.2). An unsuitable feature of administrative units is that they do not coincide with natural physiographic units in which uniform physical conditions of terrain and climate can be expected. The geographical contrasts within the boundaries of some British counties, e.g. from sea coast to mountain top in Caernarvon and Kerry, make them very unnatural units for mapping. Compare the two maps showing the British distribution of Buck's-horn Plantain, one employing Watson's vice-county units, the other a fine grid unit (Fig. 1.2). By comparison the greater scale of American landscapes nullifies this objection and maps showing distribution by counties have been used to good effect, for example in illustrating the *Flora* of the Carolinas. A further disadvantage of the ancient administrative areas of the British Isles, which does not apply in the United States, is their unequal size (generally between 200,000 and 1 million acres) and irregular shape. The rectangular counties of many American states, about the same in length, breadth and area, do not share these disadvantages and therefore do provide a very satisfactory basis for map construction, approaching in quality the grid map to which we now turn.

The most versatile system of all for mapping distributions is the grid. Its prime virtues are that the units of area are constant in size and shape and completely regular in arrangement. It is not essential that data should have been acquired by grid survey in order to justify the choice of a grid for presentation. A unique advantage of this method of plotting distributions is that the actual area of ground taken as the

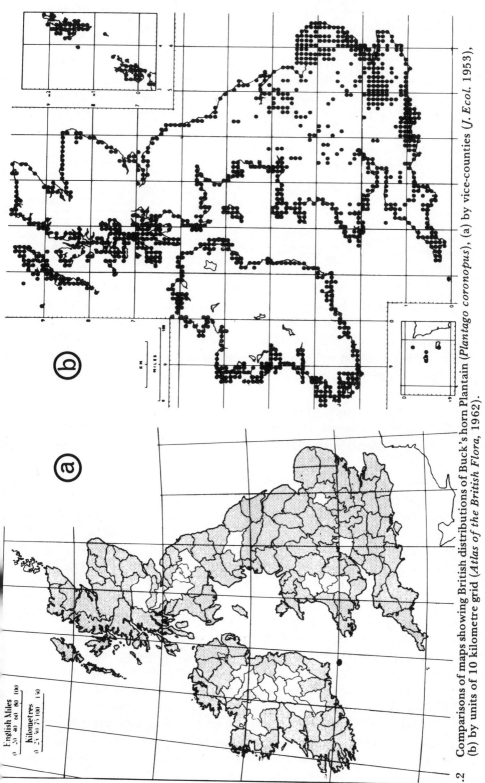

1.2   Comparisons of maps showing British distributions of Buck's horn Plantain (*Plantago coronopus*), (a) by vice-counties (*J. Ecol.* 1953), (b) by units of 10 kilometre grid (*Atlas of the British Flora*, 1962).

individual unit of the grid can be varied at will *after* data have been collected, to show whatever degree of detail is required on the finished map and to suit the scale at which it is to be published. By using a fine mesh grid in which the unit square represents something between 1 x 1km and 10 x 10km the grid map approximates to a dot map in precision. Here are some examples of maps published or in preparation which illustrate appropriate sizes of unit suitable for detailed distribution data in regions of various total areas:

	Unit Grid Square	Total Area of Territory (sq km)	No. of squares in area covered	No. of squares per 100 x 100km
Hertfordshire, England	2 x 2 km	2,000	500	2,500
Inverness, Scotland	5 x 5 km	7,500	300	400
The British Isles	10 x 10 km	350,000	3,500	100
Europe	50 x 50 km	1.5 million	6,000	4

A further benefit of the grid map is that records from several adjoining areas can be amalgamated and converted to relate to a larger unit square: indeed by use of grid co-ordinates and punched-card systems the data can be automatically re-sorted and the maps themselves produced by computer printout. To convert the distribution data for British species for incorporation in maps covering the whole of Europe every block of 25 squares on the British grid map will constitute a single unit square in the European grid. The elaboration of grid methods to cover the entire world has been described for the International Biological Programme by Perring (1967), following the GEO-Code system devised by Gould.

The basic squares of a grid can be in any convenient unit of measure. The original land survey of the United States divided the territory within each state into six-mile-square townships. Each of these townships was itself divided into 36 sections one mile square and the section corners were marked by blazes cut on trees. The modern road net of many middle western states follows a rectilinear pattern based upon these square mile sections. Here the grid is a physical reality on the ground and forms a ready-made framework for systematic survey and mapping. A most interesting series of grid maps was produced by Potzger from the records of the public land survey in Indiana. The surveyors' journals listed the location, name of species and diameter of more than 214,500 trees situated at the section and quarter-section corners and intermediate points at the time of the Indiana survey, 1799-1846. Potzger was able to prepare maps that showed not only the distribution of different tree species within the state but also their quantitative importance, using the "township" unit, a six-mile square. The extreme effectiveness of this unit for a territory as large as an entire state can be seen from the example in Fig. 1.3.

For mapping the total distribution of species it is impracticable in

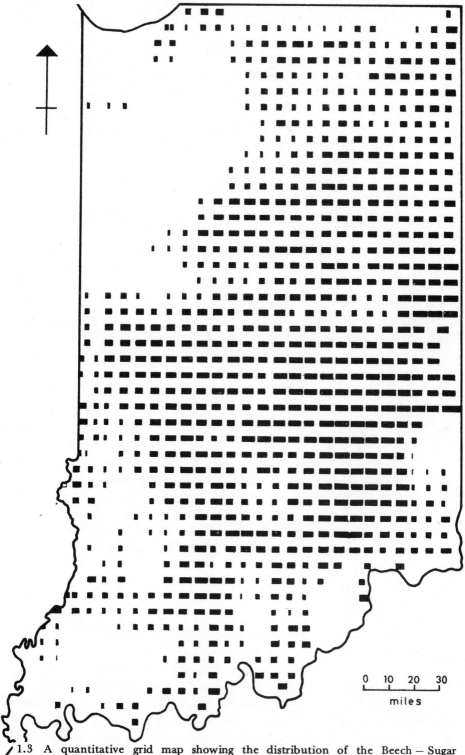

1.3 A quantitative grid map showing the distribution of the Beech – Sugar Maple – Upland Ash forest in Indiana, 1799-1846. (Redrawn from Potzger *et al* 1956).

most cases to use any of the above techniques. The quality and intensity of records varies greatly between countries so that neither dot nor grid maps can be used to full advantage. If administrative divisions are used, they will be the larger entities such as states or provinces. When those in which the species occurs have been listed, attention is usually concentrated on confirming the location of boundary stations. From the composite picture built up an outline distribution map can be drawn. It is a generalized representation often based on imperfect knowledge. Only the boundary enclosing all known areas of occurrence is shown: sometimes isolated localities beyond the general limit of the species are separately marked. The exact course of the species limit is partly dependent on the judgment of the cartographer. For the purpose of presenting a simplified picture of distribution an outline map can be produced from a dot or grid map, and if this exercise is attempted by several persons the differences between the resulting maps will clearly demonstrate the unavoidable approximation of this procedure. Maps of this type indicate that the plant or animal (or vegetation type) occurs within the area outlined but not everywhere within that area. What it states more positively is that the plant or animal does not occur outside the area outlined!

Mapping vegetation

Realizing that every species has a unique distribution and that their limits rarely coincide, we must ask "What do we mean by a vegetation boundary?" In texts describing vegetation types and the climates with which they are associated, representative examples are usually chosen from central locations within each type and the issue of boundaries or transitions is little discussed. From our rather different standpoint, in which species are considered first as the component parts from which vegetation is built, attention is focused principally on boundaries and it is a natural step from discussing the limits of species individually to considering them in their vegetational associations. The forest-prairie border in Indiana, reconstructed by Potzger from pre-settlement land survey records, provides the necessary illustration (Fig. 1.4). Across eastern and central Indiana the Beech-Maple-Upland Ash forest originally predominated in the landscape: in more western areas of the state Oak-Hickory forest was more general. However, no precise boundary can be drawn between the two forest types. There are two features to notice, since we have the records not only of the distinct associations but also of their constituent species. First, in Potzger's original account (1956) a separate map showed the relative proportions of Beech, Maple and Upland Ash individually, and it revealed that Maple began to decrease in frequency from the eastern boundary of the

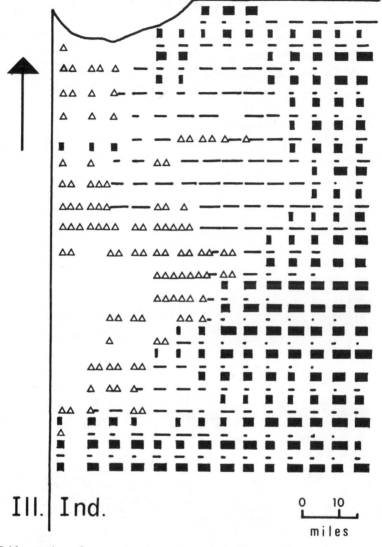

III. | **Ind.**

0   10

miles

1.4   Grid-mapping of vegetation in north-west Indiana using a six-mile square grid unit (redrawn from Potzger *et al* 1956). Thick bars (■): Beech – Sugar Maple – Upland Ash. Thin bars (—): Upland Oak. Triangles (△△△) Oak savanna. Prairie is indicated by the absence of these symbols.

state and effectively reached its own western limit before that of Beech in some areas. Upland Ash also varied in abundance independently of the other species and continued west of the limit for Beech. Therefore within a single vegetation type variations in its composition occur as the member species separately reach the limits imposed by their individual tolerances to climatic conditions. Secondly, from the map shown here (Fig. 1.4) it is apparent that Oak and Hickory species (also presented separately by Potzger) are present far into the territory that is principally occupied by Beech-Maple-Ash forest. Oak-Hickory forest first appears on warmer and drier sites on ridge-tops and south-facing slopes within the Beech forest where local conditions of aspect and exposure create a xerothermic microclimate which favours these species rather than Beech and its associates. Westward the sites occupied by Oak-Hickory gradually become more frequent and more extensive so that the transition takes the form of a forest mosaic related to land form in which the two associations change in reciprocal proportions from east to west. A comparable transition occurs in the extreme north-western counties between the Oak-Hickory forest and the prairie, through an intermediate zone of "oak-openings", i.e. oak trees in savanna formation. Towards the prairie region in the north-west part of Indiana, the hickory first diminishes in frequency leaving a forest of several species of oaks in which the trees decrease both in number and in size and a mosaic of oak woodland and oak savanna forms a ragged border with patches of tall grass prairie. The exact distribution of grass- and tree-dominated vegetation locally would be a function of micro-climate conditions related to topography. A comparable mosaic between boreal forest and tundra is illustrated in Fig. 1.5 and this is relevant to the later discussion of timberline.

Any attempt, therefore, to demarcate vegetation boundaries must be

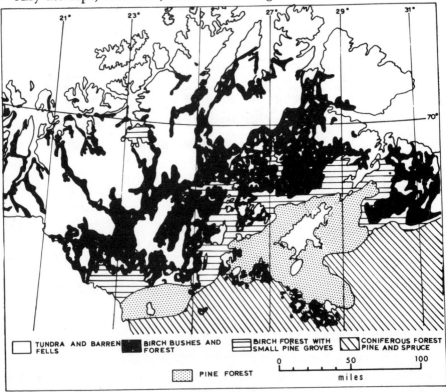

TUNDRA AND BARREN FELLS   BIRCH BUSHES AND FOREST   BIRCH FOREST WITH SMALL PINE GROVES   CONIFEROUS FOREST PINE AND SPRUCE

PINE FOREST

0   50   100
miles

1.5   The Arctic forest limit in northern Fennoscandia (after Aario, in *Atlas of Finland*, 1960). The mosaic of tundra (white) and birch scrub (black) and the irregular boundary between them reflect the influence of relief and micro-climate. Within the black zone birch varies in stature from bushes to trees: the reality of a "treeline" is elusive.

a crude simplification which ignores the transitional changes in species composition and the mosaic of stands determined by microclimate and topography within a border zone. It is excusable for practical reasons in small-scale maps, but should not be allowed to disguise the real nature of the changes represented. Even more questionable is the attempt to relate vegetation boundaries to selected climatic parameters. This is difficult enough when dealing with species individually (Chapter 4) and it must constitute a gross approximation when they are treated collectively in natural associations, if only because these associations are variable and change geographically as vegetation types are traced from place to place. A far better solution in vegetation mapping, which deserves to be more generally adopted, is the use of presence/absence records for small unit areas distributed in grid formation, precisely as used in species mapping. By this means vegetation frequency maps can be produced in which none of the plant associations have mutually exclusive areas (e.g. Fig. 3.11) and therefore the boundaries limiting the total extent of their occurrence are independent. Then, if a climatic limiting factor is correlated with the distribution of one vegetation type, it does not automatically have any implications for other types which co-exist within the same area, i.e. in a topographic and vegetation mosaic.

SOURCES OF REFERENCE

Allen, D. E. (1967) John Martyn's botanical society: a biographical analysis of the membership. *Proc. bot. Soc. Br. Isl.* 6, pp. 305-324.

Brass, L. J. (1964) Summary of the sixth Archbold expedition to New Guinea (1959). *Bull. Amer. Mus. Nat. Hist.* 127, pp. 147-215.

Brongersma, L. D. and Venema, G. F. (1962) *To the Mountains of the Stars.* Hodder & Stoughton. London. (An account of a Dutch expedition for biological collecting and anthropological study in New Guinea.)

Davis, P. H. (1965-67) *Flora of Turkey.* Edinburgh. The University Press.

Dupont, P. (1967) The Map Scheme for the French Flora. *Proc. bot. Soc. Br. Isl.* 6(4), pp. 357-61.

Futak, J. (1966) *Flora Slovenska.* The Slovak Academy.

Perring, F.H. (1967) The Geo-code. Appendix 3 in the I.B.P. Handbook No.4 compiled by G. F. Peterken. Oxford.

Perring, F. H. (1967) Mapping the Flora of Europe. *Proc. bot. Soc. Br. Isl.* 6(4), pp. 354-7.

Perring, F. H. & Walters, S. M. (1962) *Atlas of the British Flora* Text pages ix-xxiii. London & New York.

Potzger, J. E. *et al.* (1956) The forest primeval of Indiana as recorded in the original U.S. land survey . . . *Butler University Botanical Studies,* 13, pp. 95-111.

Radford, A. E., Ahles, H. E. and Bell, C.R. (1964) *Manual of the Vascular Flora of the Carolinas.* Univ. N. Carolina, Chapel Hill.

Stapf, O. (1916) A cartographic study of the southern element in the British Flora. *Proc. Linn. Soc. Lond.* (1916-17) pp. 81-91.

Stearn, W. T. (1951) Mapping the distribution of species, in "The Distribution of British Plants", proceedings of a conference of the *Bot. Soc. Br. Isl.* ed. J. E. Lousley.

Watson, H. C. (1873) *Topographical Botany.* London.

Watson, H. C. (1843) *Geographical Distribution of British Plants.* London.

CHAPTER 2

Distribution as a geographical quantity

Introduction

Clement Reid, writing in 1899, observed that climatic conditions cause two very distinct floras to be represented in Britain. The lowland flora is mainly the same as that of neighbouring lowland countries of the European continent, particularly Belgium and northern France: it is a temperate flora. The upland flora consists of numerous more or less isolated outliers of the flora that occupies the lowlands of the arctic and the mountains of Scandinavia, where it extends southwards at higher altitudes. Incidentally, the term *flora* is used to refer collectively to the complete list of plants that grow in a certain area without regard to their organisation into kinds of vegetation. These observations introduce two important biogeographic concepts. If the flora of lowland England is much like that of nearby France and Belgium, how far must you travel before the plants and animals seen are different from those in your own country? Secondly, the reference to the British upland flóra as "isolated outliers" of an arctic flora poses the question: how did these plants get to their present mountain locations? They are separated from one another by large distances and from the arctic areas of the same species by distances ten times as great. The characteristic distribution in Britain of mountain species comprises a number of discrete areas whose elevation is sufficient to provide the appropriate climatic conditions. To understand how such a distribution came into being it is necessary to accept one of two alternative hypotheses, either that the dispersal capacity of plants is unlimited or that the distributions were at one time continuous and have subsequently been restricted, i.e. environmental conditions have changed. The first theory assumes that wherever suitable conditions exist for the growth of a particular species the plants are capable of reaching the locality and establishing themselves in the appropriate habitat. If this suggestion is correct, distance is not regarded as a barrier to effective dispersal and no delay is inferred providing that seed is available in an already flourishing population. However, the diversity of size, weight and form of plant seeds and the variety of dispersal mechanism in different species suggests that the dispersal range cannot be the same for all. Indeed, even the smallest air-borne seeds may be carried only a few miles in the air currents of velocities usually experienced. If appeal is made to exceptional circumstances occurring at infrequent intervals then a probability factor is introduced and long periods of time may be necessary for the rare occurrence of extremely long-range dispersal. In general it is doubtful that any plants have really long-distance dispersal without the aid of special agencies such as transport by particular species of birds, by rivers or by ocean currents. Even in these special cases the dispersal will be strongly directional and will not enable the plant to reach all situations where it could grow. For the majority of plant species whose seeds are not transported by these means i.e. not eaten by birds and which grow in situations not accessible to transport by water (e.g. plants of mountain tops), the distance seeds are dispersed is extremely limited and of a scale more related to ensuring survival of the population within the locality than to colonizing other areas. Praeger (1911) measured the rate of fall in still air of seeds of 90 species representing many families and including most of the variations in seed size and shape that occur in the Irish flora. While some plumed seeds had up to ten times the buoyancy of un-plumed seeds, the latter, though small, have a high specific gravity and even minute examples of what he called "powder seeds" recorded rates of fall between 4ft and 10ft per second.

The alternative theory is that geographical, climatic and ecological changes have in the course of time altered very drastically the conditions of a landscape in which a species originally spread so that occupation of areas now isolated from each other was achieved when suitable conditions extended without interruption between them. Edward Forbes (1846) explained the presence of arctic plants in Britain, "stranded on our mountain peaks", by referring to an earlier time when a similar flora covered the whole of the British Isles and was not confined to these isolated localities. This view can be considered more fully at a later stage when other contributory information has been discussed.

Patterns of distribution in small territories

A more detailed classification of the species that populate a given territory involves distinguishing several geographical categories according to the areal bias shown in individual distribution maps. Species which occupy the same parts of the territory or whose distributions show similar tendencies, even though their boundaries are not coincident, can be assigned to the same group and referred to as a

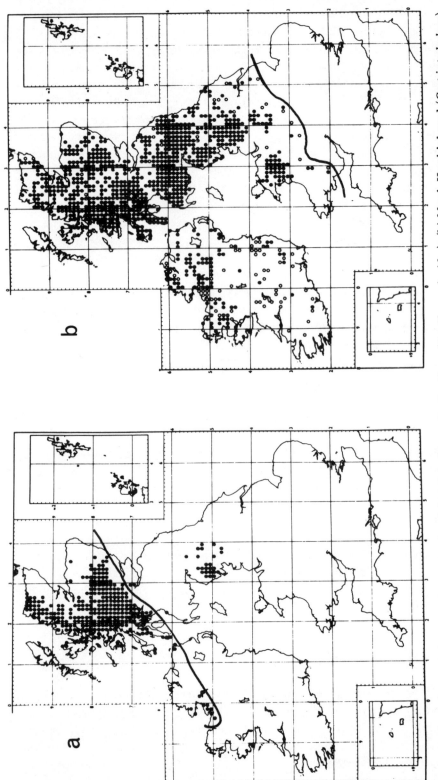

2.1   Northern distributions in the British Isles: (a) Yellow Mountain Saxifrage (*Saxifraga aizoides*), (b) Marsh Hawk's-beard (*Crepis paludosa*). (*Atlas of the British Flora*, 1962).

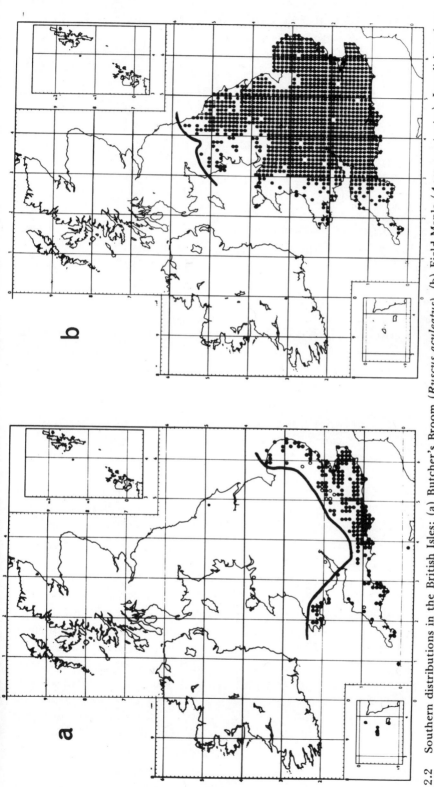

2.2 Southern distributions in the British Isles: (a) Butcher's Broom (*Ruscus aculeatus*), (b) Field Maple (*Acer campestre*). Localities where the plants are introduced (i.e. not native) are omitted. Maps kindly supplied by Biological Records Centre, Nature Conservancy, Huntingdon.

geographical element. Just how different the geographical distributions of species that occur in the same country can be is illustrated in Figs. 2.1-2.5 with reference to British plants. Within the country a species may be distinctly northern, southern, eastern or western: even south-eastern and south-western distributions may be distinguishable. This is essentially the kind of consideration that geographical elements are founded upon. The examples show that species belonging to the same element, those with northern distributions for instance, extend very different distances to reach their limits (towards the south in this case), cf. Yellow Saxifrage, Trailing Azalea and Hawk's-beard (Fig. 2.1), see also Figs 4.4-4.5. Similarly the distribution of southern species (Fig. 2.2) range in area from extremely southern (e.g. Butcher's Broom) to those whose limit of distribution is far to the north (e.g. Field Maple). Many species with distributions of intermediate extent could be mentioned (e.g. Wayfaring Tree and Old Man's Beard) and it is clear that the member species of each group recognised by a particular geographical bias can therefore be arranged in a series. A further point to notice is that the northern and southern groups do not occupy mutually exclusive territories. Many members of the contrasted groups have distributions that overlap towards their respective limits, e.g. Hawk's-Beard and Field Maple.

It happens that the examples selected to illustrate eastern and western distributions, i.e. Crosswort *(Galium cruciata)* and English Stonecrop *(Sedum anglicum),* have almost exclusive areas but a narrow zone of overlap can be detected between their areas of frequent occurrence, bounded by the heavy lines in Fig. 2.3. Scattered occurrences of each species transgress still further into the territory of the other. These particular plants, however, do not compete with one another since they require entirely distinct habitats and are therefore segregated ecologically even where their distributions coincide. Competition cannot be the reason for their geographical segregation and climatic influences are probably the cause.

A south-western element is distinguished from generally western species like English Stonecrop by the fact that their distributions show a northern limit within the British Isles as well as an eastern limit. Again the constituent species can be arranged in a series showing "equiformal progressive areas" (Hulten, 1937), that is, their patterns of distribution are essentially similar but expand to differing extents. Thus Tree Mallow *(Lavatera arborea)* extends slightly further to the north around the Irish Sea (Fig. 2.4) than Wild Madder *(Rubia peregrina)* which extends slightly further to the east along the English Channel. Does this reflect slight differences in the nature of the climatic factors influencing them? It would be easy to describe such distributions as "penetrating" to differing extents but this might suggest that they are immigrating or have immigrated in a north-easterly direction (from where?). Is this a

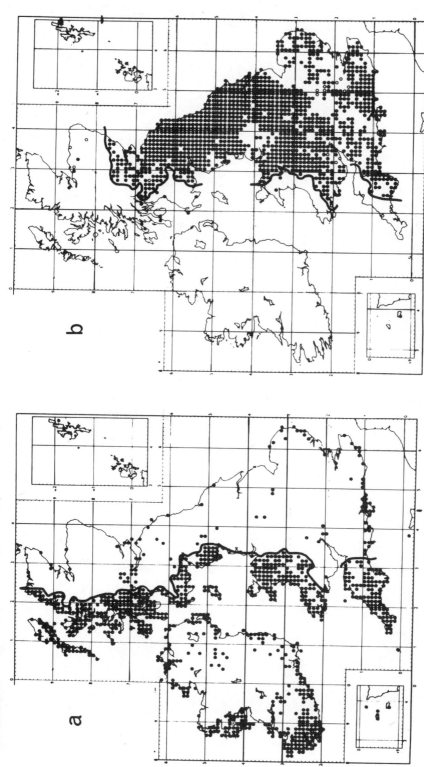

2.3   Western and eastern distributions in the British Isles. (a) English Stonecrop (*Sedum anglicum*), (b) Crosswort (*Galium cruciata*). (*Atlas of the British Flora*, 1962.)

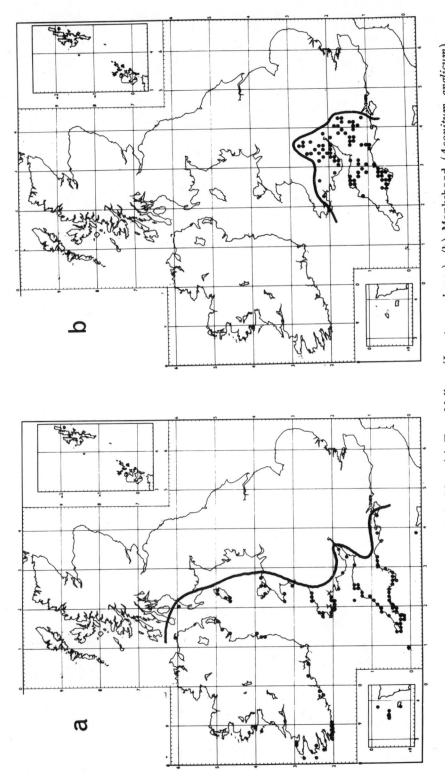

2.4   South-western distributions in the British Isles. (a) Tree Mallow (*Lavatera arborea*), (b) Monkshood (*Aconitum anglicum*). Introductions omitted. Maps kindly supplied by Biological Records Centre, Nature Conservancy, Huntingdon.

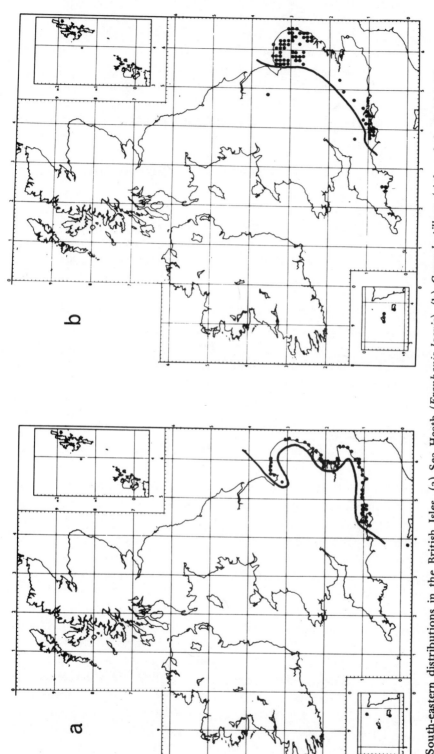

2.5   South-eastern distributions in the British Isles. (a) Sea Heath (*Frankenia laevis*), (b) .*Crassula tillaea*. (*Atlas of the British Flora,* 1962.)

justifiable inference? If we consider the geographical characteristics of plant or animal species, our attention is focused on fundamental questions such as these. The south-western plants already discussed are both found on cliffs, the mallow being exclusively coastal. Another species belonging to the same geographical element but occurring in an entirely different habitat (woods and shaded situations) is Monkshood (*Aconitum anglicum*) and its distribution includes localities much further inland, though still in the south-western counties. South-eastern distributions are illustrated by the Sea Heath (*Frankenia laevis*) which is a coastal plant and by *Crassula tillaea* (Fig. 2.5), which is discussed in a later section (p. 46).

It might be possible to separate yet another geographical element from the great variety of British distributions. Mountain plants found only at high elevations might be recognised separately but there are few mountain plants that do not occur, anomalously perhaps, at low elevations in some places (e.g. western Ireland) and all of them are quite appropriately included in a general northern element.

Watson (1873) recognised six broad elements within the British flora. The first of these, called the British type, is inevitably a large group and is not very informative since it includes all species that occur widely throughout England, Wales, and Scotland, i.e. not showing any geographical limit to their distribution within Britain. The remaining five elements are distinguished by only partial occupation of the country as follows:

English type: occurring chiefly in England, especially in the south, and becoming rarer northwards. *Scottish* type: occuring chiefly in Scotland or reaching the north of England, becoming scarce southwards. *Highland* type: restricted to mountain areas. *Germanic* type: restricted to east and south-east England. *Atlantic* type: found chiefly in west and south-west Britain.

Disadvantages can be seen, however, when this scheme is compared with a classification that takes into account not merely the distribution within Britain (or other comparatively limited areas) but the full extent of geographical range in Europe as a whole. One practical limitation of Watson's scheme is that so many species reveal no geographical bias at all because they are found throughout Britain: it does not follow that they are equally widespread throughout Europe as a whole. For example, two heath species, Cross-leaved Heath (*Erica tetralix*) and Bell Heather (*Erica cinerea*), occur throughout Britain wherever appropriate habitats exist, but to an observer in Poland they would be unfamiliar as they are absent or local in that country (Fig. 2.8). The same is true of Gorse (*Ulex europaeus*), Foxglove (*Digitalis purpurea*), Bluebell (*Endymion non-scriptus*) and many other plants of diverse ecological situations that are so common in Britain that one might regard them as characteristic plants of our landscape. To the observer in central Europe these plants are west European and can be associated with the oceanic climate which influences those regions adjoining the Atlantic coasts. This is an indication that the whole of the British Isles lies within "Atlantic" Europe. Although a climatic gradient can be recognised from western Ireland to East Anglia even a species demanding an equable climate will not necessarily find its particular limit of tolerance within this range. Compared with the climate of Poland, that of East Anglia is oceanic. In attempting to characterize the climatic affinity of any plant, therefore, it is preferable to examine its distribution over a large territory (continental or sub-continental) so that its relation to a fairly extreme range of climatic conditions may be represented by its area of occurrence.

Patterns of distribution at continental scale

In the geography of the flora and fauna of Europe several patterns of distribution frequently recur and obviously these bring the plants and animals they represent into geographical relationship. Distributions that are centred in the same region and whose boundaries extend in broadly the same directions can therefore be grouped together and regarded as members of the same geographical element. Of course, the size and extent of the areas of member species vary and their boundaries often do not coincide: their individual limits may be spread over a zone covering several degrees of latitude or longitude, but the essential similarity of areal form and location indicates their geographical alliance.

About ten such elements can be recognised in Europe as well as the category of widely distributed species, that is to say species whose range is so extensive within the continent, both from north to south and from east to west, that no regional label can be attached to them. Many of these extend their distribution beyond the boundaries of Europe into Asia and if we were adopting a world basis for classification would be named within a system of larger categories as for example "European – West Siberian". However, for the purpose of a European classification the extra-European distribution of a species is not considered. Some, though by no means all, of the species included as regional elements in this scheme also occur in other parts of the northern hemisphere. This is particularly true of many arctic, boreal and northern Atlantic species, which in many cases have even larger areas of occurrence in North America or in northern U.S.S.R.

The scheme described and illustrated here (Figs. 2.6-2.13) is original and is intended to demonstrate the principles which underlie the more elaborate classifications of other authors (e.g. Good, Hultén, Matthews,

Linnaea borealis

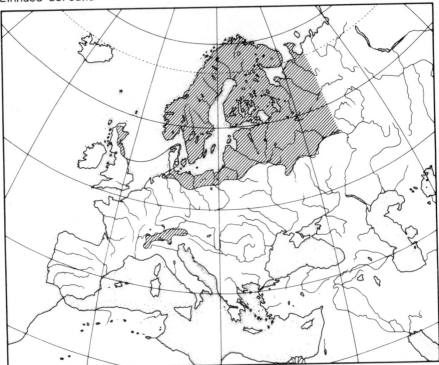

Cymindis vaporariorum

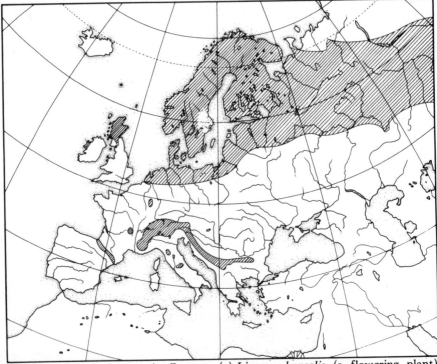

2.6 The Boreal element in Europe. (a) *Linnaea borealis* (a flowering plant), (b) *Cymindis vaporariorum* (a carabid beetle). After Matthews 1955 and Deville 1930.

Lobelia dortmanna

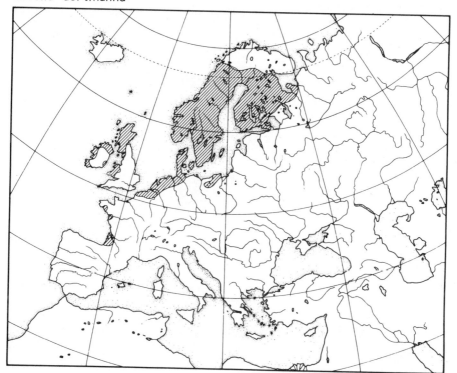

Pterostichus adstrictus

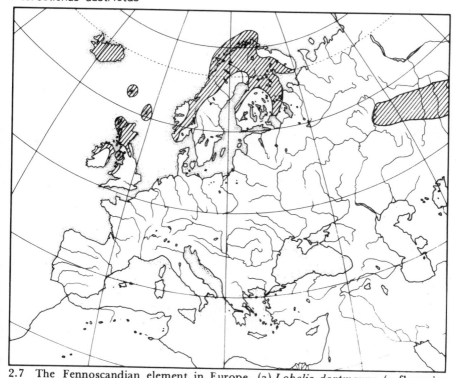

2.7 The Fennoscandian element in Europe. (a) *Lobelia dortmanna* (a flowering plant), (b) *Pterostichus adstrictus* (a carabid beetle). After Matthews 1955 and Deville 1930.

Erica tetralix

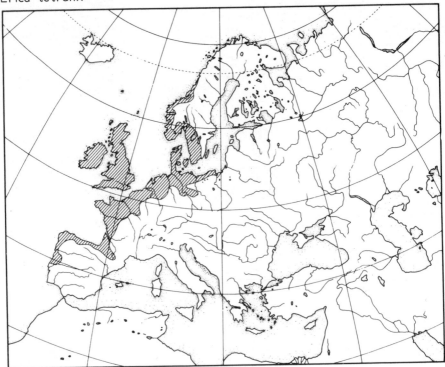

Myrica gale

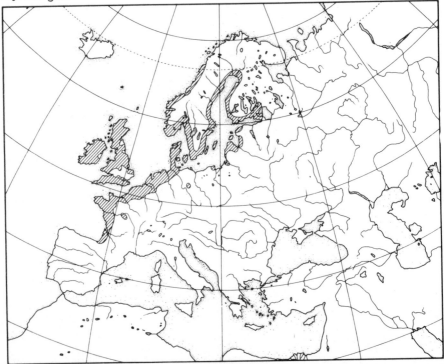

2.8 The Eu-atlantic element in Europe. (a) *Erica tetralix*, (b) *Myrica gale* (both flowering plants). After Walter 1954 and Meusel *et al* 1965.

Lonicera periclymenum

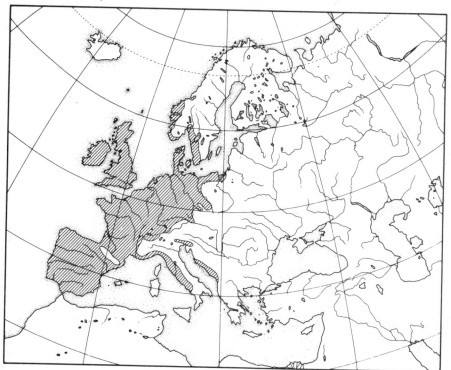

Nebrix degenerata

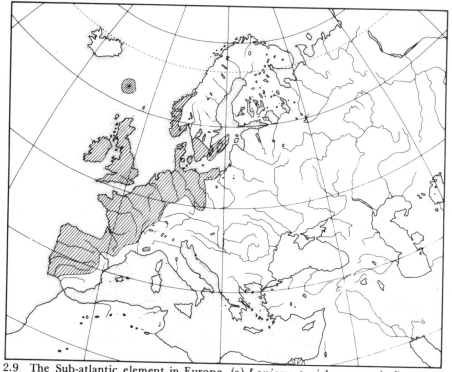

2.9 The Sub-atlantic element in Europe. (a) *Lonicera periclymenum* (a flowering plant), *Nebrix degenerata* (a carabid beetle). After Walter 1954 and Deville 1930.

Meusel, Walter). It may be compared with the American schemes of Dice (1943) or Gleason and Cronquist (1964).

The concept of geographical elements relates to all kinds of plants and animals and its usefulness is strengthened in that the elements recognised, say, for flowering plants, can be directly applied to lower plants such as mosses and liverworts (Hepatics), and to invertebrate animals such as snails and insects, as some of the maps included here demonstrate.

The European elements can be named as follows:

1. *Boreal Element.* Species which occur within the area of northern forests of predominantly coniferous composition but not necessarily in forest habitats. Distributions normally occupy a broad band in high latitudes and extend eastwards often undiminished up to and beyond the Ural Mountains. Equivalent to Matthews' "Continental Northern". Examples: *Andromeda polifolia, Linnaea borealis* (Fig. 2.6).

2. *Arctic-alpine Element.* Species whose main distribution is in the arctic latitudes, usually extensive at low elevations but including mountain areas and extending south into lower latitudes at high elevations. Often with isolated areas in mountains distant from the arctic (Alps, Carpathians, Pyrenees). Examples: *Loiseleuria procumbens* (Trailing Azalea), *Dryas octopetala,* (Mountain Avens) (Fig. 10.16).

3. *Fennoscandian Element.* Species which have their distribution centred in the Scandinavian peninsula including Finland, usually with an eastern limit in Europe near the White Sea. Includes many of Matthews' "Oceanic Northern" species but not those whose Scandinavian distribution is confined to western coasts (see 4 below). Examples: *Primula farinosa* (Bird's-eye Primrose), *Lobelia dortmanna,* (Water Lobelia) (Fig. 2.7).

4. *Eu-atlantic Element.* Species whose distribution is greatly extended in latitude but is always approximate to Atlantic coasts and their narrow hinterland. At their northern and southern extremities they very narrowly fringe the coasts of Scandinavia or Iberia. Some are found only along the more northern coasts and were included by Matthews in his "Oceanic Northern" element while others found mainly along the French coasts and southward were included in his "Oceanic West European". Both types overlap in the British Isles and intergrade and therefore no attempt is made here to differentiate on a north-south basis. Examples: *Erica tetralix* (Cross-leaved Heath), *Myrica gale* (Bog Myrtle) (Fig. 2.8).

5. *Sub-atlantic Element.* Species whose distribution is west European, penetrating further from Atlantic coasts than the Eu-atlantic type and generally including a large territory in Iberia or in Europe west of Jutland. Their northern limit in some cases reaches Scandinavia

where they are resticted to a narrow coastal fringe. Corresponds to part of Matthews' "Oceanic West European" except for those species which qualify as Eu-atlantic.

Examples: *Ilex aquifolium* (Holly), *Digitalis purpurea* (Foxglove), *Lonicera periclymenum* (Honeysuckle) (Fig. 2.9).

6. *Iberian Element*. Species with south-western distribution in Europe and widespread in the Iberian peninsula. These do not reach North Sea coasts but in many cases occur throughout the western Mediterranean. This includes many of Matthews' "Oceanic Southern" species but others included by him in the same group, whose range is incomplete in Iberia and which reach the North Sea area, fall naturally in the Sub-atlantic type (see 5 above).

Examples: *Rubia peregrina* (Wild Madder) (Fig. 2.10), *Erica scoparia* (Fig. 10.10).

7. *Mediterranean Element*. Species occuring in areas surrounding the Mediterranean Sea, usually including parts of the North African region but excluding the central areas of the Iberian and Balkan peninsulas. In most cases freedom from frost is an important factor in determining their northern limits.

Examples: *Arbutus unedo* (Strawberry Tree), *Lavatera arborea* (Tree Mallow), *Frankenia laevis* (Sea Heath), *Erica arborea* (Tree Heath) (Fig. 2.11).

8. *South European Element*. Species of wide distribution in Europe south of the Pyrenean, Alpine and Carpathian mountain chains belong in this group. Generally widespread in the Iberian, Italian and Balkan peninsulas, in many cases they include also the Pontic region (Black Sea). Some species reach northward into the Pannonian and Bohemian basins of Hungary and Czechoslovakia.

Examples: *Quercus pubescens* (Fig. 2.11).

9. *Balkan Element*. Species centred in the south-east of Europe find their largest territory in the Balkan peninsula for which this group is named. Many of them also occur around the Black Sea (Pontic) and in the Caucasian region while some extend northwards as far as Bohemia.

Examples: *Loranthus europaeus, Carpinus orientalis* (Fig. 2.12).

10. *Nemoral Element*. Species having a wide distribution in the latitudes naturally occupied by deciduous forest to the exclusion of coniferous forest (after the definition of Sjörs 1963). The eastern limit in this type is often a narrowing tongue extending towards the southern end of the Ural Mountains but in other cases runs from the southern Baltic Sea towards the Black Sea and may extend to include the Caucasus. The species of this element do not usually penetrate beyond northern Iberia (Cantabrian Mountains) and only in some cases does the northern limit reach southern Sweden.

Examples: *Viscum album* (Mistletoe), *Corylus avellana* (Hazel), *Ulmus glabra* (Elm) (Fig. 2.13).

Rubia peregrina

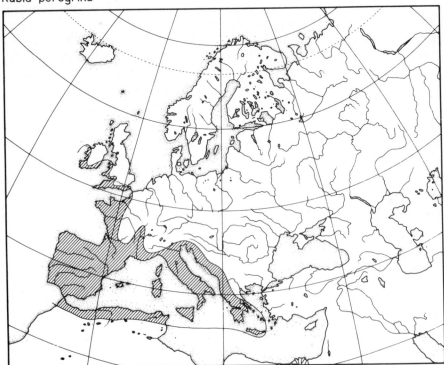

Notiophilus quadripunctatus

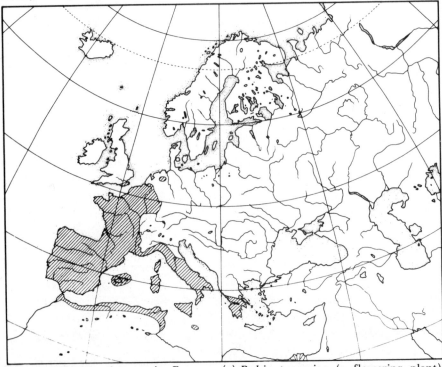

2.10 The Iberian element in Europe. (a) *Rubia peregrina* (a flowering plant), (b) *Notiophilus quadripunctatus* (a carabid beetle). After Matthews 1955 and Deville 1930.

Erica arborea

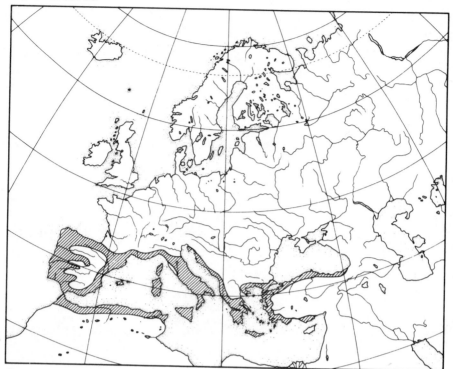

Quercus pubescens

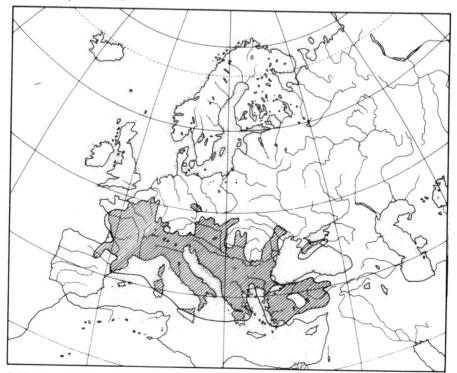

2.11 The Mediterranean and South European elements. (a) *Erica arborea*, (b) *Quercus pubescens*. After Rikli 1943 and Meusel *et al* 1965.

Loranthus europaeus

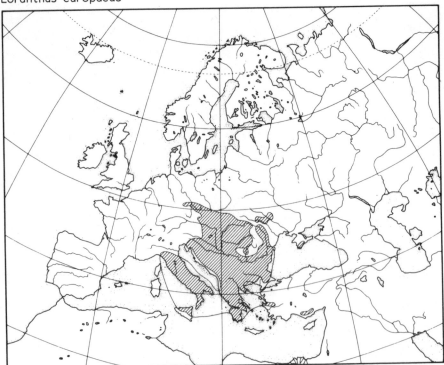

Carpinus orientalis

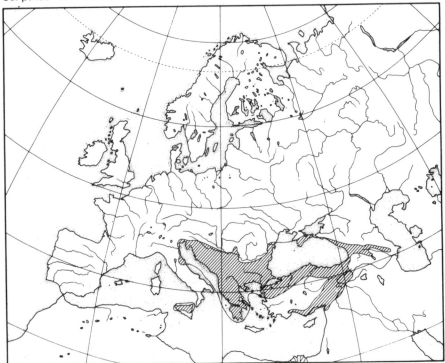

2.12 The Balkan element in Europe. (a) *Loranthus europaeus*, (b) *Carpinus orientalis*. (both flowering plants) After Meusel *et al* 1965.

Viscum album

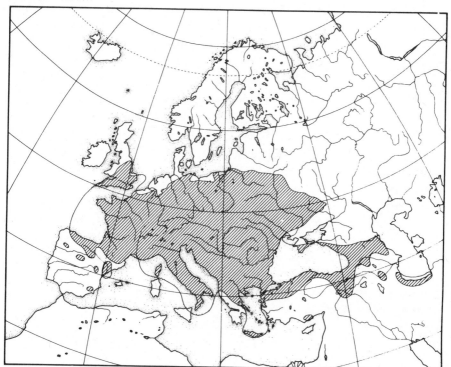

Corylus avellana

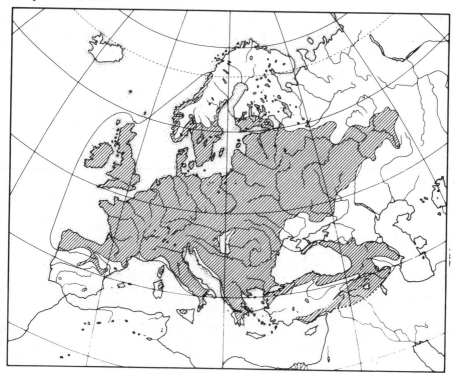

2.13 The Nemoral element in Europe. (a) Mistletoe (*Viscum album*), (b) Hazel (*Corylus avellana*). After Meusel *et al* 1965.

11. *European Wide Element.* Species whose range is more extensive than the last (10), i.e. reaching to between 40° and 60° E longitude and penetrating further north in Scandinavia and the Baltic region.
Examples: *Pinus sylvestris* (Scots Pine).

The only plants and animals which cannot be accommodated in this scheme are those whose distribution is of the type called *narrow endemic,* that is, their total area of occurrence is confined to just one region or physiographic unit such as a plateau, mountain chain, (even an isolated peak), an island or group of islands, a river basin, etc. There are examples of species with very restricted distribution in the Balkan highlands and others which are solely found in the Pyrenees, the Alps or Carpathians. For further discussions of such cases see Chapter 9.

Any classification may be designed with one of two distinct objectives in mind. The intention may be to demonstrate essential similarities between organisms or areas, i.e. the classification is built upon the concept of *relationship.* Alternatively, the intention may be to discriminate between groups of organisms or areas as far as possible, i.e. the classification directs attention to features that *distinguish* rather than to those which *relate.* The results of the two treatments are quite different and it is clearly the one based on relationship that is appropriate to the recognition of geographical elements.

Since geographical elements can be used as a tool for the analysis and description of the plant and animal life of a state or country it is worth considering some implications and defects of using the alternative discriminatory kind of classification. As an example, the scheme produced by Deville as a basis for discussing the European ranges of beetles is cited.

Coleoptera of the British Isles grouped according to their European distributions (St. Claire Deville, 1930).

Group A	Centred in the middle of the Eurasiatic continent extending westward in a progressively narrowing tongue. Many of them do not reach the Atlantic and those which do so border it only on a fairly narrow front.
Group B	Species which outside the British Isles are only found in the other Northern Atlantic Islands (Iceland, Faroes), in Scandinavia and in arctic regions.
Group C	Species which outside the British Islands inhabit not only Scandinavia and arctic regions, but also the high mountains of temperate Europe and even of southern Europe.
Group D	Species whose area of distribution extends to the islands of the Mediterranean and to North Africa, or at least to one of these regions.

	Subdivision (i)	Eurytherm insects, present in parts of Denmark, northern Germany, Scandinavia and the Baltic States.
	Subdivision (ii)	Species that do not extend beyond the south coast of England (c. 50°N lat).
Group E		Species not reaching or not passing beyond the north coast of the Mediterranean at their southern limit.
	Subdivision (i)	Eurytherm species of wide distribution.
	Subdivision (ii)	more markedly atlantic and usually limited outside the British Isles to more southern latitudes.

Particular attention is given here to the extremities of many distributions, a feature that is shared to some extent by the geographical classification of European plants by Matthews (1955). Deville's Group B is distinguished from Group C essentially by whether or not the species occur in the mountain ranges of temperate latitudes in Europe. Such montane areas in any species are small in comparison with their extensive distribution in the arctic (especially in Scandinavia) and in the latter respect both groups are identical. To separate them on this minor criterion is artificial because disjunct areas are notable examples of the effect of historical factors and in the light of fossil evidence the location and indeed the existence of any disjunct area is seen to be a consequence of chance factors. An identical criterion is used by Matthews to distinguish "Continental Northern" plants from those classed as "Northern Montane". In the scheme presented here these species are all included in the Boreal element which simply recognises the area in which their most continuous and extensive distribution is achieved.

Use of the furthest north or furthest south occurrence to define other groups is exemplified by Deville. Group D requires that species reach *southward* to the islands of the Mediterranean or to North Africa while Group E includes those whose *southern* boundaries stop at the northern, i.e. European, shore of the Mediterranean. Both groups are sub-divided according to the position of the *northern* limit of distribution, either extending into North Europe or not present north of the English Channel coast (c. latitude 50°N). Clearly such distinctions are arbitrary and tend to separate species whose areal pattern is very similar in form. There is also a strong risk that by confining attention solely to these points of a particular nature the gross features of distribution may be overlooked and that species whose general patterns of area are very different may be included in the same group on these very limited criteria. The resulting classes are undoubtedly artificial, in contrast to the more natural classification previously described. It is not claimed that there are no difficulties in the natural classification. There are all kinds of intermediate distributions which form series grading imperceptibly from one element to another, but a subjective decision has to be taken in only the most transitional cases, which are a small minority. Where rigid demarcation lines are laid down, as in artificial classifications, a large number of species are allocated by such divisive criteria. It follows that a much greater proportion of species will be placed in satisfactory, i.e. natural, geographical alliance by avoiding divisive procedures.

Clustered and dispersed distributions

A problem of geographical classification based upon a rather small territory is that species present in only a few scattered localities or condensed into a very small part of that territory may be difficult to assign with certainty to any of the named groups. Watson recognised this problem by creating two special categories; the "Local" type for sparse and scattered distributions in which the localities of occurrence are highly dispersed, e.g. Nottingham Catchfly *(Silene nutans)* (Fig. 2.14), and the "Intermediate" type for more compact distributions occupying geographical positions which would be inappropriately included in either "English" or "Scottish" groups, e.g. Baneberry *(Actaea spicata)*, Bog Rosemary *(Andromeda polifolia)*, Bird's-eye Primrose *(Primula farinosa)* (Fig. 2.15). Watson used the "Intermediate" group particularly for plants which occur in limited areas between southern Scotland and northern England but under another name this group could be enlarged to include many other species whose *form* of distribution (not geographical position) shows the same characteristics, i.e. the plant is present at quite numerous localities within an area of small total extent. The pattern of distribution in such cases is clustered. The problem with distributions of this kind is that although the location of each cluster is known the broader geographical tendency of these species cannot be diagnosed from such small areas. For example, if Bird's-eye Primrose is to be regarded as a northern species in Britain its absence from the entire area of Scotland north of the central lowlands is inconsistent. Similarly if St. Dabeoc's Heath, which occurs only in County Galway, is considered with other western species in Ireland its absence from other Atlantic peninsulas is enigmatic. If we limit our consideration of species such as these to the evidence of their distribution within the British Isles then neither climatic nor geographical affinities can be safely inferred. However, if we take into account their occurrence elsewhere in Europe we find that they have disjunct distributions,* of which the small areas in the British Isles are just isolated parts.

Disjunct distributions are those in which parts occur in two or more regions separated by relatively great distances. The disjunct areas may be on the same continent or on different continents.

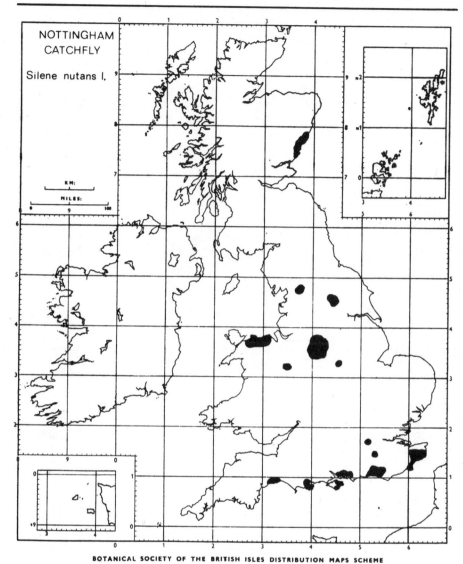

2.14 Nottingham Catchfly (*Silene nutans*): a dispersed distribution. (Redrawn from *Atlas of the British Flora* 1962.)

Watson's "Local" type represents a kind of distribution that is discontinuous or hyper-dispersed. Individual localities of the plant's occurrence are separated by distances too great to be explained by colonization of one locality from another by natural means of dispersal. In this respect it contrasts with the clustered type of distribution in which the localities of occurrence tend to be grouped within each territory, even to the extent that interchange between local populations is possible; but such areas are separated by distances or barriers of a magnitude that excludes the possibility of dispersal from one to another. Special considerations apply to the interpretation of disjunct and discontinuous distributions which we will discuss later (Chapter 9).

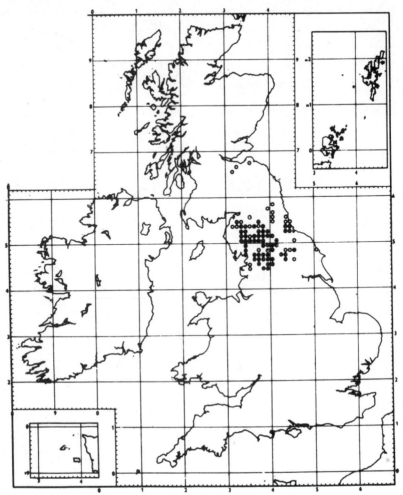

2.15 Bird's-eye Primrose (*Primula farinosa*): a clustered distribution (*Atlas of the British Flora*, 1962.)

Comparison of regional and continental distributions

Another problem arises when the category to which a species is assigned on the basis of its distribution within a region does not correspond with the element that is indicated by its general distribution at continental scale. To demonstrate this difficulty, consider the distributions in Britain of the Stonecrop *Crassula tillaea,* a tiny succulent-leaved annual plant growing in sandy and gravelly places, and the related Rock Stonecrop *(Sedum forsteranum)* which grows on cliff-ledges and in rock crevices. Their distinct distributions in Britain are respectively south-eastern and south-western (Fig. 2.16). It would be reasonable to expect that *Sedum forsteranum* is distributed in western regions of Europe south of the English Channel and equally reasonable to expect that *Crassula tillaea* extends southward and eastward into France, i.e. representing a "continental" element in the British flora. In fact, while

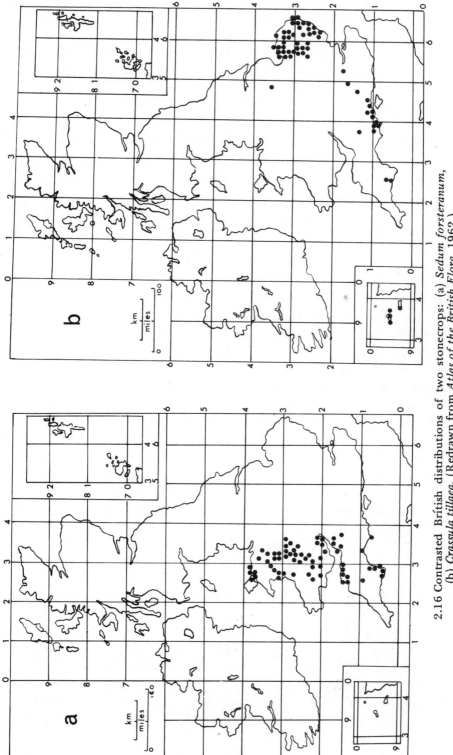

2.16 Contrasted British distributions of two stonecrops: (a) *Sedum forsteranum*, (b) *Crassula tillaea*. (Redrawn from *Atlas of the British Flora*. 1962.)

the *Sedum* conforms to expectation in possessing an Atlantic distribution (Oceanic West European of Matthews), the *Crassula* is a west
Mediterranean species (Oceanic Southern of Matthews) and therefore
departs considerably from the expected pattern. The sandy and gravelly
habitats required by *Crassula* are common in southern and eastern
England on soft sedimentary rocks and superficial deposits of Tertiary
and Quaternary age, but are absent, for the most part, on the hard
Palaeozoic rocks of the south-west. The higher summer temperatures of
the south-east may also perhaps be necessary to this plant, here at the
northern limit of its range. Thus, *Crassula tillaea* is considered an
"oceanic" plant in Europe but it does not occupy the most oceanic
parts of the British Isles for reasons determined by its other
requirements.

Classification of resident flora and fauna according to their distribution within a single region or other relatively small study area may
be strongly influenced by peculiar circumstances of geology, geomorphology (relief, slope, dissection), soils and by the history of that
particular landscape. For the same reasons, species with narrow
endemic and disjunct distributions should be excluded when treating
geographical elements at continental scale.

The erroneous impression gained by Watson, from study of only
British distribution areas, that species strictly limited to south-western
districts were necessarily of Iberian or Lusitanian affinity and that
those of the south-east were Germanic or Teutonic, was also shared by
nineteenth-century entomologists in Britain. Their conclusions were
mistaken only because they did not investigate the wider range of the
species outside the British Isles (Deville, 1930). Their interpretations
were particularly suspect when they assumed that the distributions
reflected the directions from which the species had immigrated. The
extreme case is that of two or more species having the same
distributions in England but whose European distributions, when
studied in their entirety, reveal that they belong to quite different
elements. The ground-beetles *Agonum lirens* Gyllh, and *Licinus
punctatulus* F. *(granulatus* Dej) are both restricted to the extreme
south-east of England yet the first has a nemoral distribution in Europe
and the second is Iberian (Fig. 2.17). Secondary factors are responsible
for anomalous local distribution.

The general case may be stated thus: at or near the geographical limit
of a species its occurrence may be controlled by edaphic or micro-
climatic requirements and therefore be associated with particular
geological and topographic situations where these requirements are met.
Its marginal distribution may not entirely coincide with the places
where the gross climatic conditions normally associated with the species
are most fully developed. Due to the restricted terms of reference of
state or regional classification, species may be assigned to the same

Agonum lirens

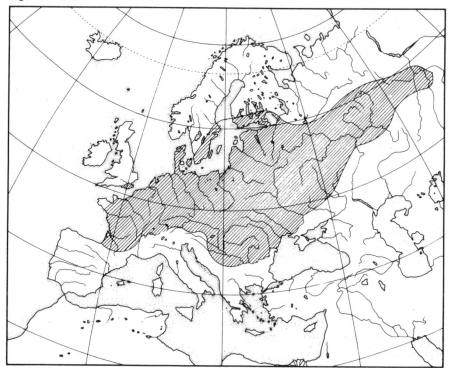

Licinus punctatulus

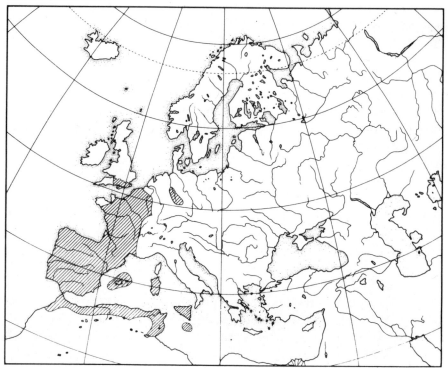

2.17 Contrasted European distributions of two ground-beetles. Their areas in Britain are coincident *viz.* south-eastern. After Deville 1930.

group whose distributions are approximately coincident within the district (but not outside it) for reasons determined by quite disparate factors. Probably any group defined on a limited areal basis should be considered simply as a descriptive unit, broadly reflecting the present climatic (perhaps micro-climatic), physiographic or edaphic association of those plants and animals whose distribution shows a common local pattern.

The term geographical element should be reserved for categories based on total distribution within areas of at least continental size. Species allied geographically in this way have a long-standing association due to the more widely pervasive influence of climate.

The historical perspective

The importance of knowing something about the history of particular plants and animals and environmental conditions should be apparent. It is remarkable, therefore, that Edward Forbes himself adopted a historical explanation of the elements he recognised in his classic memoir (1846). His view was that the different geographical components were the result of distinct migrations into the British Isles, each under different climatic conditions, at different times and from different source areas and that the influence of climate in earlier times is more important in explaining the original entry of the species than the present climate which is merely responsible for maintaining them in the areas they now occupy. It is particularly relevant to explaining discontinuous distributions, e.g. of montane species, for in these cases there seems to be no chance under present circumstances for plants to spread from one suitable locality to another distant locality which also possesses suitable conditions for their growth. The proposition that geographical elements are migrational assemblies in this historical sense does not dissociate them from climatic implications but rather strengthens these implications and adds a new dimension to the interpretation of distribution — that of time.

The idea that every species has its own history in terms of area occupied is undoubtedly correct but the historical explanation offered by Forbes has its difficulties. Forbes suggested that the Lusitanian element (and the Atlantic element) were probably of pre-glacial origin in Ireland and south-west Britain, i.e. the immigration of these two groups took place sometime before the Pleistocene glaciation and that since their invasion they have persisted in the British Isles. No doubt his own experience as a member of the Geological Survey had made him familiar with the fossil flora of certain Tertiary rocks in south-east England, notably the Eocene London Clay, and the sub-tropical character of this more ancient vegetation (see Chapter 10) had indicated to him that climates of greater warmth than at present were to be expected before the glacial period rather than after it.

The difficulties arise when it seems necessary to postulate conflicting climates (at different times) to account for the immigration of different elements of the flora. Thus Forbes believed that the arctic-alpine element migrated into Britain during the glacial period, a view that was very far-sighted at a time barely ten years after the theory of glaciation had been put forward. It is still regarded as eminently reasonable and indeed it is now substantiated by abundant fossil evidence. However, this thesis was later challenged by Clement Reid (also of the Geological Survey) in 1899 on the argument that if plants of southern origin, thought to be warmth-requiring (thermophilous), did enter the British Isles before the glacial period, how did they then survive the prolonged cold episode in which great ice-sheets extended from the north to cover much of the British Isles? Without going into the ramifications of a controversy which raged in learned journals for another three decades, we can see the fundamental issues raised by two alternative inter-pretations of geographical elements. There are obvious merits in both views and both of them also present some problems.

In one view geographical elements are seen as groups of species with generally similar climatic tolerance which have spread to occupy approximately coincident areas limited by present climatic control. This approach is concerned mainly with the *status quo* and carries no implications about the times at which the distributions were achieved. Different times could conceivably apply to the various members of the same element, allowing for different rates of dispersal among species. Nor is there any implicit assumption that members of the same group have spread initially from the same source areas. They are linked only by their occupation of a common territory. In the historical view the geographical element is an entity in its past as well as in the present. Its member species are thought to comprise a migrational group which, if they did not actually spread simultaneously, entered as waves of migration within a single climatic period which provided appropriate conditions.

Deville expressed the view that for beetles the influence of climate has been greatly exaggerated and that it is only a decisive factor in certain well-defined examples. The most important factor for these insects is the physical texture of the ground surface and the character of the soil. He interpreted geographical elements in the beetle fauna as historically associated groups whose species migrated by the same routes, contained by physiographic barriers and facilitated by natural corridors which existed formerly even if no longer operative. Illus-trating the concept of migrational episodes related to particular climatic periods, he believed that one element (of the British coleopteran fauna) arrived by penetrating from the east in conditions of a steppe-like character (as indicated by their ecology) before the British land connection with Europe was inundated by marine transgression. Then the sea barrier was created and by the following reasoning he deduced that only after this event did climatic conditions become suitable for forest growth. Recognising a European element of beetle species which occur in forest habitats from central Europe to northern France, e.g. Pas de Calais, but not represented in the British Isles, he suggested that these species migrated over forested territory at some time after the straits now separating England from France became flooded. On this view their distribution is related rather to a particular period of time in the postglacial history of Europe, namely, after the formation of a sea barrier, than to any present operation of climatic factors.

Biogeographers of three and four decades ago did not always appreciate that even if elements do represent migrational assemblages – which is a matter open to question – their present distributions may not disclose the directions from which they first entered the territory under discussion. It is implicit in the migrational theory that species within the same geographical element spread from the same source region. Time is indeed a potent factor and any immigrant group may suffer many vicissitudes in the course of its subsequent history under the influence of climatic changes, through the competitive effects of later invaders and by more gradual and persistent changes in the soil and land surfaces as a result of prolonged weathering. The present extent of the territory occupied by an element and by its individual species is almost certainly not the same as the area it first colonized on entry: it may be greater or it may well be smaller. Particularly if it has undergone reduction in area it may have been displaced so that what remains is merely a fringe of its original distribution. The origin of fringe elements such as the more extreme Eu-atlantic species now confined to the western seaboard of Ireland and to the south-western peninsulas of England and Wales must be considered in this light. The truth may lie in a combination of circumstances and both theories may be partly correct: past events and present conditions exert controlling influence on the distribution of living things.

SOURCES OF REFERENCE

Deville, J. St. Claire (1930) Quelques aspects du peuplement des Iles britanniques. (Coléoptères) *Mémoires de la Société de Biogéographie*, vol. 3, pp. 99-150. Paris.

Forbes, E. (1846) On the connection between the distribution of the existing fauna and flora of the British Isles, and the geological changes which have affected their area . . . *Memoirs of the Geological Survey of Great Britain*, 1, 336-432.

Matthews, J.R. (1955) *Origin and Distribution of the British Flora*. Hutchinson, London.

Praeger, R.L. (1911) Biological Survey of Clare Island: Phanerogamia and Pteridophyta. *Proc. Roy. Irish Acad.* vol. 31, part 10.

Reid, C. (1899) *Origin of the British Flora.* Dulau & Co., London.

Turrill, W.B. (1948) *British Plant Life.* Collins, London.

Watson, H.C. (1883) *Topographical Botany.* 2nd Ed. London.

Watson, H.C. (1847-59) *Compendium of the Cybele Britannica* or British plants in their geographical relations. Longmans, London.

CHAPTER 3

The influence of local factors on distribution

However widespread the distribution of a species, it may in fact be present in very low density on the ground and may not be among the species that you would notice as "abundant" or "common" within any one locality. To understand the relationship between the local occurrence of plants and animals and their total area of distribution we must look more closely at the details of distribution within a small part of their total range. In doing so, the importance of knowing something about the ecology of species becomes evident: indeed the geographical study of species over small areas reveals much of the circumstantial evidence of ecology. By examining the location of individuals and of colonies or communities in relation to the physical features and conditions of the terrain we can make out the controlling influences that govern the pattern of their occurrence in the territory at large.

In Britain there is a long tradition of detailed studies, particularly of plants, within the relatively small areas of individual counties, which are not too large for a naturalist to get to know intimately during a decade or a lifetime of study. At first the objective was simply to record by name the places where the plant was seen but since the emergence of ecological thought, recording has extended to description of the type of situation occupied at each locality so that ultimately habitats in which the plant finds conditions within its tolerance can be listed. By their frequency in this list the habitats most favourable to the species are identified and with this information the observer would know where to expect the same species to be present within a different and unfamiliar area. Furthermore the mapping of distribution in detail over small areas, such as counties, provides evidence of the density of populations in different species and allows this to be correlated with features of local physiography, geology, soils and climate.

Habitat data were prominent in the work of Horwood (1933) on the flora of Leicestershire and Rutland and the production of maps to express the ecological and topographic behaviour of plant species was a prime objective in *A Geographical Handbook of the Dorset Flora* by Good (1948), from which the following examples are taken.

Situated in the south of England, Dorset has a coastline of some 70 miles but the straight-line distance from west to east is only about 42 miles and from north to south the county extends 24 miles. It is therefore comparable in size to many American counties but within this small area it contains unusual topographic and geological diversity (Fig. 3.2). In the east is a synclinal basin in which Tertiary rocks are preserved, diagonally across the county from north-east to south-west a broad outcrop of Mesozoic chalk is exposed in the form of an escarpment and in the west and north-west there is a complex of vales exposing lithologically varied strata of lower Mesozoic age. The vales present a marked relief which accentuates the geological variety in this way: all the principal elevations are capped by deposits of Upper Greensand lying unconformably on Liassic formations, chiefly clays and marls, which are exposed on the lower slopes and bottomland. Thus the county as a whole has a particularly wide range of environments within its limited expanse.

Effects of climate

Unlike changes in rock outcrop from place to place, which show abrupt boundaries, the changes in climate within so small a territory are not extreme and take the form of gradients. We can therefore expect that while rock or soil characteristics may determine sharp limits to plant distributions where strongly contrasting substrates adjoin, climatic influence may show itself by altering the frequency of suitable habitats and hence of local populations. For example rainfall lies within the range 43″ to 27″ annual average, and is greatest on the high ground in the west and north, lowest along the south coast and around Poole Harbour in the east. In the maps for some species Good detects a corresponding gradient in the frequency of localities at which the plants are present. The dot distribution of Harts Tongue Fern is dense in the wetter districts of west and north and sparse or absent in the driest parts of the county (Fig. 3.1). In such a topographically varied landscape it would be unrealistic to say that its distribution is limited by a certain critical value of annual rainfall. All that has been demonstrated is a correlation, not a causal relationship.

It is true that this plant requires a rather moist habitat and absence of drought but these conditions are not met everywhere within the area of higher rainfall — nor excluded from some locations receiving lower rainfall. The conditions experienced by the plant are those of the micro-climate in the places where it is growing, i.e. in this example measurable within one foot of soil surface. The moisture supply as soil water (rather than direct rainfall), the absence of strong insolation and consequent evaporation and the local persistence of highly humid air,

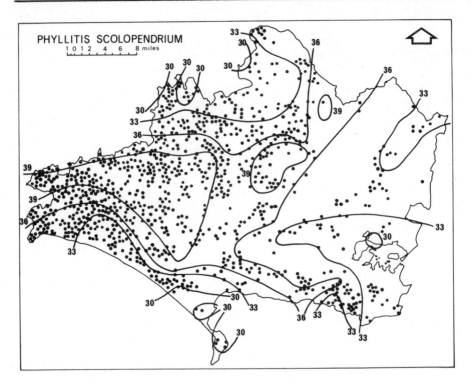

3.1   Dorset: distribution of Harts Tongue Fern and mean annual rainfall (in inches). Note gradient in density of plant distribution from west to east. (Redrawn from Good 1948.)

e.g. as on a clay stream-bank, are the actually effective factors. The correlation with rainfall distribution merely indicates the greater probability of such habitats existing in higher rainfall districts, subject to there being suitable terrain to contribute the other requirements. Closer inspection of the map shows that, despite the general trend, the fern is missing from part of the high rainfall area in the north-east, which is an exposed plateau undissected by streams, and is present in some low rainfall areas near the south coast where numerous ravines and more humid air from the sea compensate for diminished total precipitation. This example illustrates the indirect link between the plant's micro-climate and that described by standard climatological records. Correlations may be detected but in most cases these reveal only the probability of certain micro-climatic conditions being encountered.

Effects of substrate

Many of the Dorset plant maps show patterns of distribution that can be correlated principally with the geological map or with the soil map. The influence of climate, if apparent at all, is superimposed on to this

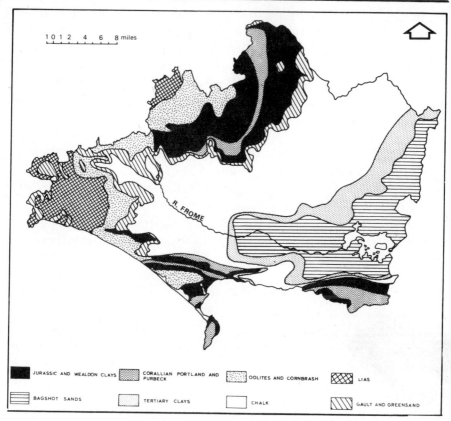

JURASSIC AND WEALDON CLAYS　　CORALLIAN PORTLAND AND PURBECK　　OOLITES AND CORNBRASH　　LIAS

BAGSHOT SANDS　　TERTIARY CLAYS　　CHALK　　GAULT AND GREENSAND

3.2　Dorset: solid geology. (Redrawn from Good 1948.)

pattern and is not so easily detected. The correlations with types of rock and soil are extremely interesting: for the ecologist they confirm what he has learned from the *intensive* observation of the plants, and place this knowledge in a geographical context; and to others the maps, carefully interpreted, convey the essence of the plants' ecology and relate their occurrence to topography. It is perhaps worth remarking that ecologists who rely wholly upon experiments under controlled conditions and on casual acquaintance with their object of study in the field are often unaware of the subtle indications of ecological tolerance that can be revealed by detailed distribution maps and by *extensive* observation over a wide territory. This is the method of biogeography.

　　Showing very close correlation with the occurrence of calcareous soils is the distribution of Hoary Plantain (*Plantago media*) (Fig. 3.3). Its localities are all areas where the bedrock is chalk or limestone and so the map betrays very clearly this aspect of its ecological requirements. Another plant that shares the same preference for lime-rich soils (such plants are termed calcicole) is Old Man's Beard (*Clematis vitalba*) and while its distribution in Dorset (Fig. 3.4) also relates to the outcrop of the same rock types it is noticeably different in detail from that of the Plantain. In particular it is remarkably scarce on the chalk in the valley of the Frome and to the south of this river in the central part of the county. Some other influence must be responsible for its absence in this district, an indication that although its essential requirement for calcareous soils is satisfied, other conditions are also necessary to constitute a tolerable habitat and some of these are not fulfilled in the southern part of Dorset. *Clematis* is a climbing plant that relies upon other woody species for support and it is therefore usually found in woods and hedges. The relative scarcity of woods in this part of the county may in part provide an answer to this question. This brings us to a general point, that even species with important ecological require-ments in common do not have identical distributions because of the differential effect of other secondary factors. Even without the influence of contributory factors, the distribution patterns of species that show ecological affinity — frequently occurring in association — are never the same. Instead of being coincident their distributions have a tendency to fall into series of progressively increasing area. The existence of such series indicates the operation of an environmental gradient, that is, a continuous scale of values for the quantity of any physical variable such as the capacity of the soil to retain moisture under drought or its acidity or alkalinity. Further examples will demonstrate this.

The common Heather or Ling (*Calluna vulgaris*) is confined to soils with free acidity (such plants are termed calcifuge or acidophile). It is tolerant of a range of drainage conditions from very wet bog soils to very porous sandy 'heath' soils so long as its prime requirement is met. In Dorset (Fig. 3.5) this species is almost confined to the Tertiary basin in the east where both the Bagshot Sand and the London Clay, though differing in drainage properties, are deficient in basic ions and therefore acid. Outside this district the occurrences of heather are very scattered because suitable habitats exist only exceptionally on other geological formations where leaching of the soil has progressed to the stage at which acidity develops in the upper horizons. It is possible to find heather locally even on the chalk areas where superficial deposits of loessic origin have been leached of carbonates.

In the heathlands around Poole Harbour another plant frequently associated with heather on the well-drained acid soils is Bracken (*Pteridium aquilinum*) (Fig. 3.6), a fern which is found with only minor variation in many parts of the world. Thus we may be sure that it is tolerant of acid soils but its distribution in Dorset is not confined to the places where heather grows. It occurs not only in the Tertiary basin on podsolic soils but also in the western part of the county on brown earths and other non-calcareous soils. Its wider area of distribution reflects wider range of tolerance.

There is no sharp line of distinction between acidic and alkaline soils but in reality a continuous series of intermediate conditions exists, all

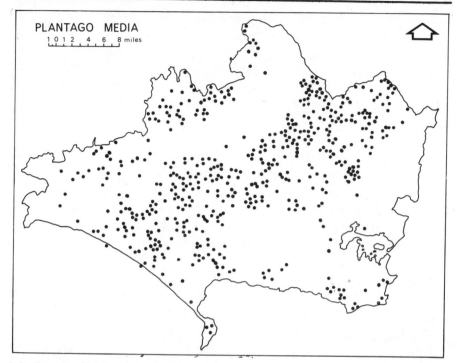

3.3 Dorset: distribution of Hoary Plantain (redrawn from Good 1948) – a plant of lime-rich soils.

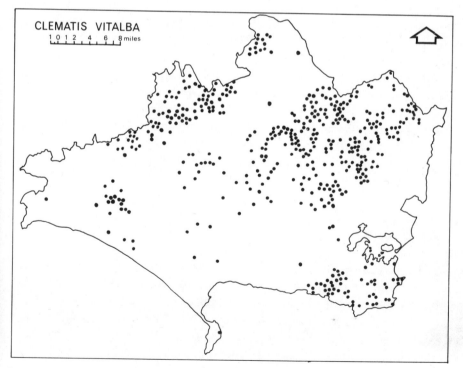

3.4 Dorset: distribution of Old Man's Beard (redrawn from Good 1948) – a plant of lime-rich soils.

of which are measurable on the scale of hydrogen-ion concentration. As different plant species have different ranges of tolerance on this scale, not mutually exclusive but extending to different limits, it follows that this will be expressed in their areal distribution since the conditions to which they respond are associated with changes in relief, drainage, soil types, etc. All studies of small territories, if carried out in sufficient detail, reveal relationships of this kind. The distribution areas of various species can be sorted into groups, all members of each group showing similar patterns (though not identical as explained above), those of one group contrasting with those of other groups. The species with distributional affinities do not necessarily share the same habitats but some common factor between their habitats is indicated and they are certainly related to the same facets of the local landscape.

Another English county comparable in geology and in relief to Dorset is the county of Hertford situated a hundred miles to the north-east. In a detailed study by Dony (1967) a grid system was used for survey and also for mapping. In the resulting maps, instead of the point localities being marked by the smallest possible dot that is legible at the scale chosen (as in the Dorset examples), all individual localities within each grid unit of 2 x 2 km are registered as a single large dot, the size of which on the map is purely conventional. Nevertheless, the unit is still sufficiently small to make detailed interpretation possible. This county also is crossed by the Cretaceous chalk escarpment from north-east to south-west though this is covered by superficial deposits of calcareous glacial boulder-clay in the north and by residual clay-with-flints, which produces acidic soils to the south of the glacial limit (Fig. 3.7). Thus, contrasted and intermediate soil types are present here too and the distribution of various species conforms to their behaviour as described in Dorset. The corroborative evidence from two counties makes us more confident that mapped distributions over small areas can truly reflect the ecological demands of plant species and especially their tolerance with regard to soil characteristics, collectively known as edaphic factors.

Edaphic tolerance

In Hertfordshire the association of certain species with particular soil types is well brought out in the distribution maps. Heather, which is a plant of distinctly acid soils, has a rather limited number of occurrences in this county and these are mainly within the districts of Quaternary gravel deposits and clay-with-flints. A series of maps (Fig. 3.8) shows other species whose range of tolerance is more or less restricted to acidic soils. Wood Sage is a plant which often occurs within the regions of gravels and London Clay but hardly ever on other substrates

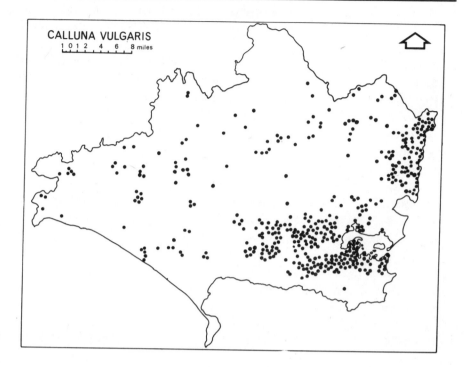

3.5 Dorset: distribution of Heather — a plant of lime-deficient soils. (Redrawn from Good 1948.)

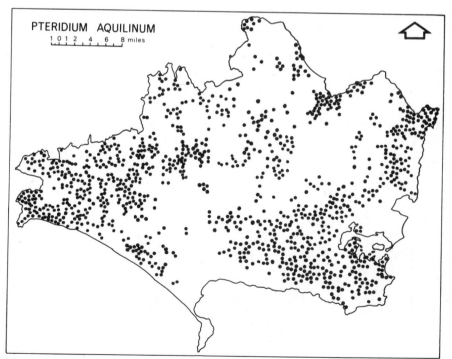

3.6 Dorset: distribution of Bracken — a plant of well-drained soils. (Redrawn from Good 1948.)

in Hertford. Foxglove also occupies similar habitats and occurs in the same districts but in addition appears in woodlands on the clay-with-flints where soil acidity is not so extreme. Bracken is present in all these areas and also extends into the boulder clay district while the final member in this series, Honeysuckle, includes the whole of the boulder clay area within its distribution and is only absent from the northern border of the county where raw chalk soils occur.

The frequency of these plants increases in that order, showing that they find suitable habitats at a greater number of places, and they also show progressive increase in area of distribution within the county. Both these features indicate greater range of tolerance to edaphic conditions. Thus Bracken has greater tolerance than Foxglove which in turn has greater tolerance than Wood Sage. In fact in this series tolerance can be referred to a graded series of soils ranging from podsolic to acid brown earth to neutral brown earth. The final examples such as Holly and Honeysuckle are tolerant of a wide range of soils, excluding only those containing free lime. The distribution of such plants is complementary to that of calcicolous species e.g. Horseshoe Vetch *(Hippocrepis comosa)*, which grow only where lime-rich soils occur, cf. maps 3.8d and 3.8h. The calcicoles also form a series in which every species displays a different limit to its tolerance by the extent of its distribution on the map. In this case tolerance relates to diminishing

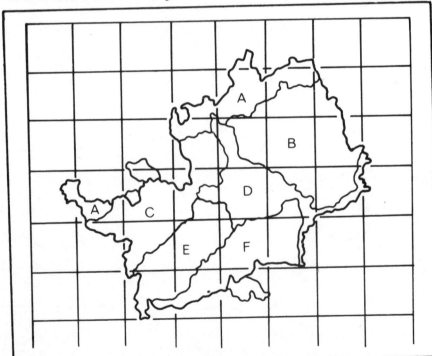

3.7   County of Hertford: geological areas. (After Dony 1967) A — chalk escarp-ment; B — boulder-clay (till); C — clay-with-flints; D — and E — flint gravels; F — London Clay.

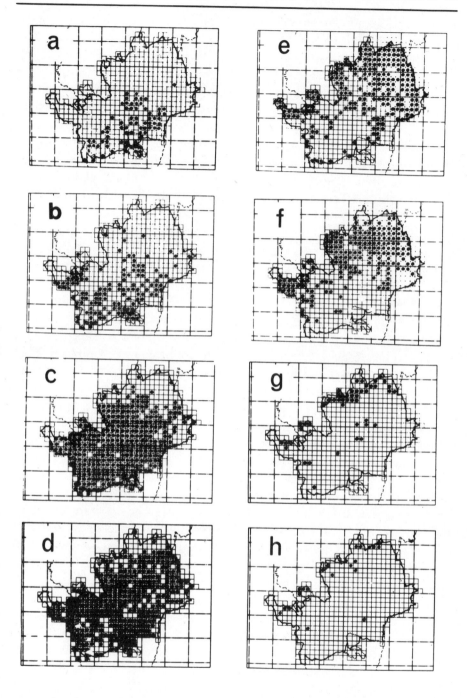

3.8 County of Hertford: selected plant distributions, from Dony 1967. (a) to (d): a calcifuge series, (e) to (h): a calcicole series. (a) Wood Sage (*Teucrium scorodonia*), (b) Foxglove (*Digitalis purpurea*), (c) Bracken (*Pteridium aquilinum*) (d) Honeysuckle (*Lonicera periclymenum*), (e) Cowslip (*Primula veris*), (f) Wayfaring Tree (*Viburnum lantana*), (g) Kidney Vetch (*Anthyllis vulneraria*), (h) Horseshoe Vetch (*Hippocrepis comosa*).

concentrations of lime and changes in the availability of other nutrients associated with this factor.

Of the species illustrated, most exacting in its requirements is the Horseshoe Vetch, followed by the Kidney Vetch. Their distributions in Hertfordshire are confined to the outcrops of chalk along the escarpment itself and to very few other places, possibly where chalk bedrock has been exposed by quarrying (Fig. 2.8 g-h). Small Scabious (*Scabiosa columbaria*) also has a similar distribution. The more extensive distribution of the Wayfaring Tree (*Viburnum lantana*) shows that it is tolerant of soils on the chalky boulder clay and on clay-with-flints where this is not completely decalcified, as well as on the chalk itself. The same tendency is carried a stage further in the distribution of Cowslip (Fig. 3.8e). As a wide-ranging plant of at least somewhat calcareous soils its distribution excludes that of Wood Sage, an obligate calcifuge (acidophile) species. (Fig. 3.8a). Of course every small square of these maps represents an area of 4 sq km and naturally there are soil variations related to topography, even where the same substrate provides a uniform basis for soil formation. Consequently the same square may provide records for both a moderate calcicole and a moderate calcifuge. That is why the distributions of Wayfaring Tree (*Viburnum lantana*) and Foxglove (*Digitalis purpurea*) coincide in some squares. However, extremely contrasted soil types rarely occur within so short a distance (2 x 2 km) and therefore the ecological differences between species are generally revealed at this scale of mapping, especially those relating to edaphic conditions.

At smaller scales the control of species distributions by edaphic factors is still apparent in the case of plants with strict requirements, i.e. rather narrow ecological range, such as those demanding calcareous soils. Indeed many plants of chalk and limestone habitats could be arranged in order of decreasing range of tolerance (ascending order of demand for and dependence on lime) simply by careful comparison of detailed grid maps of their distribution in the British Isles (cf. Perring and Walters, 1962). The most widespread of these calcicolous species e.g. Kidney Vetch and Common Rockrose faithfully repeat in their distribution the pattern of chalk and limestone outcrops across the country, and also include some areas of shelly lime-rich sands and other superficial deposits such as glacial drift where these incorporate material derived from calcareous bedrock. If the map for Rockrose (Fig. 3.9) was shown to a geologist, he would not fail to recognise the distribution and would probably be able to name the rock formations represented. Calcicoles of more limited tolerance, e.g. Small Scabious and Squinancy Wort, are more strictly confined to the outcrops of the purer limestones and the chalk (Fig. 3.9). However, while the influence of soil types and the underlying rock formations dominates the distribution of plants such as these, the effects of climatic influences also become

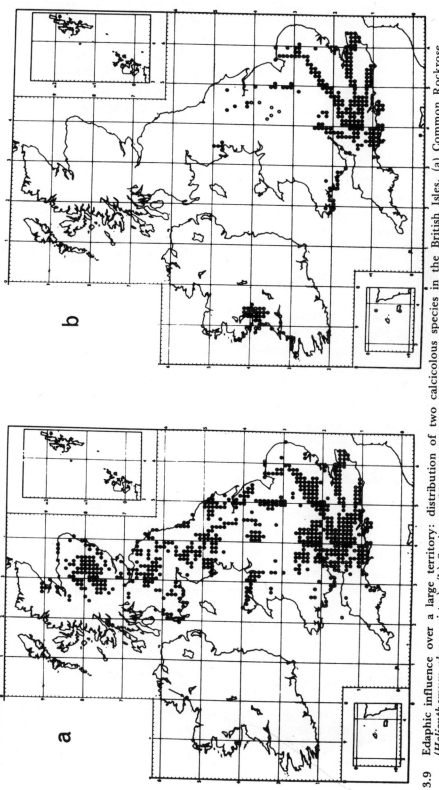

3.9   Edaphic influence over a large territory: distribution of two calcicolous species in the British Isles, (a) Common Rockrose (*Helianthemum chamaecistus*), (b) Squinancy Wort (*Asperula cynanchica*). (*Atlas of the British Flora*, 1962.)

noticeable at the scale of national or state maps. Thus Common Rockrose *(Helianthemum chamaecistus)* occurs in diminishing frequency towards the west and is almost absent in western Scotland and south-west Wales even where limestones are exposed. It is notably absent from the considerable tracts of Carboniferous limestone in Ireland which seem to supply exactly the same physical conditions as the corresponding habitats in Britain. In the case of Squinancy Wort *(Asperula cynanchica)* climatic influence is presumably responsible for its diminishing frequency northward and for its complete absence from apparently suitable habitats in north Wales, northern England and the whole of Scotland.

Ecological segregation

While discussing how the influence of local factors may be detected in distribution maps, the example of montane plants deserves notice. As with plants of the lowlands, species with different soil preferences can be distinguished and indeed have been listed separately after extensive study by McVean and Ratcliffe (1962). These authors divide the mountain flora into five categories. For each of these, tolerance limits have been determined by chemical analysis of soils from many representative stands of vegetation (Table 1).

Table 1:   **Plant tolerance of some edaphic factors.**

	Soil Calcium as mg/100 gm dry soil	Acidity (pH)
Exacting Calcicoles	> 300 mg	>6.0 (usually)
Calcicoles	> 30 mg	> 4.8 (usually)
Calcifuge	< 30 mg	< 4.5
Species avoiding poorest soils	> 15 mg	> 4.5

The species which "avoid" the poorest soils are absent from the most lime-deficient and highly acid substrates but do encroach into part of the tolerance range of calcifuges and therefore overlap into some typically calcifuge associations. The fifth category is named "indifferent" and is necessary to account for a considerable group of plants which can occur over an extremely wide range of soil calcium-status and acidity. However, they may usually be associated with a restricted part of this range and occur only sparingly on soils at one extreme or the other.

   Cowberry *(Vaccinium vitis-idaea)* and Bearberry *(Arctostaphylos uva-ursi)* are both dwarf shrubs classed as indifferent in regard to their soil requirements. They grow in pine woodland, moorland and mountain heath at altitudes within the forest, subalpine and low alpine zones in Scotland, i.e. up to 3,500 ft and much the greatest area of

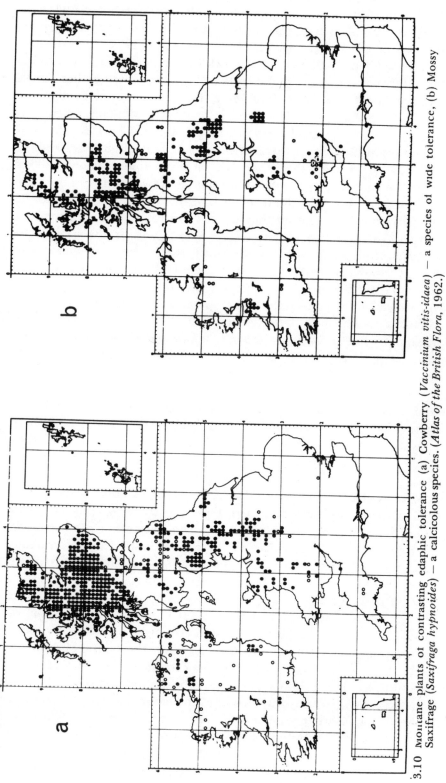

3.10 Mo11a11e plants of contrasting edaphic tolerance (a) Cowberry (*Vaccinium vitis-idaea*) – a species of wide tolerance, (b) Mossy Saxifrage (*Saxifraga hypnoides*) – a calcicolous species. (*Atlas of the British Flora*, 1962.)

these vegetation types occurs on acidic and base-deficient soils. Valley-slopes, plateaux, ridges and summits are the topographic situations in which these species are found. From the range of habitats available to them a wide distribution in Britain can be expected and in general terms this is realized (Fig. 3.10). However, Cowberry has a more extensive distribution than Bearberry, being widespread in northern Britain and occurring in Wales and in the north Midlands where it is present in heaths at an elevation of only 500 ft (152 m). Bearberry is uncommon outside the Scottish Highlands and although it descends to 200 ft (61 m) within that region it is never found below 1000 ft (305 m) in England. Thus while both species grow together in the same plant communities throughout much of their range and conform to the same pattern of distribution they differ in the extent of their areas. In other words, there are districts from which one or the other (but usually *Arctostaphylos*) is absent despite the presence of appropriate habitats, suitable soils (for it is not demanding) and the usual associate species. This difference is superimposed, as it were, on fundamentally similar distributions and reflects the operation of some climatic factor.

Let us compare with these two a group of calcicole montane plants, for example Mossy Saxifrage (*Saxifraga hypnoides*), Roseroot (*Sedum rosea*), Alpine Chickweed (*Cerastium alpinum*), Alpine Cinquefoil (*Potentilla crantzii*) and Mountain Avens (*Dryas octopetala*). These may be grouped geographically by virtue of their similar distributions and ecologically because they occupy much the same habitats and grow as direct associates in many cases. Their habitats are distinct from those of Cowberry and Bearberry and are separately located even when they occur on the same mountain but of course such local segregation does not register on small-scale maps. In general the distribution maps of these five species, with minor individual differences, contrast with those of Cowberry and Bearberry and thus express geographically their distinct habitat preferences (Fig. 3.10). The calcicoles named grow on cliff ledges and in rock crevices on outcrops of base-rich and calcareous strata, not only limestones, but lime-bearing intrusive and extrusive igneous rocks, e.g. pumice tuffs, dolerite, gabbro, as well as certain schists. These plants are therefore to be found mainly where these rocks are exposed and where soils derived from them have their supply of nutrient elements constantly renewed through irrigation by lime-charged water. Thus the cliffs themselves and the talus slopes below them, where seepage water emerges, and stream gullies on these rock types provide the habitats and the topographic situations for the calcicoles. The different distribution of the moorlands characterized by Cowberry and Bearberry is shown in the maps of the vegetation types (Fig. 3.11). The distribution of each vegetational association in which a species frequently occurs is of course less widespread than that of the species considered in isolation. The vegetation units recognised by

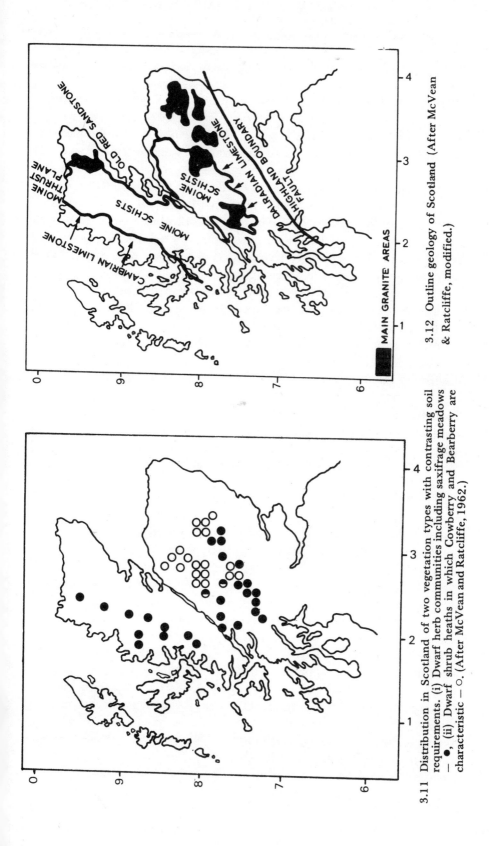

3.11 Distribution in Scotland of two vegetation types with contrasting soil requirements. (i) Dwarf herb communities including saxifrage meadows — ●, (ii) Dwarf shrub heaths in which Cowberry and Bearberry are characteristic — ○. (After McVean and Ratcliffe, 1962.)

3.12 Outline geology of Scotland (After McVean & Ratcliffe, modified.)

McVean and Ratcliffe depend upon the concurrence of several species in intimate association, i.e. within representative stands of a few square metres. In the "dwarf herb communities" of these authors, including those designated "saxifrage meadows", all five of the species under discussion occur together and the distribution of the vegetation types thus constituted is shown in Fig. 3.11. It corresponds with the outcrops of various lime-bearing rocks of different ages and origins (Fig. 3.12). Included are Dalradian limestones exposed in a broad band trending from south-west to north-east between the mountain districts of Breadalbane and Clova in east-central Scotland, and calcareous mica-schists of the Moine and Lewisian series north of the Great Glen. The distribution of the "dwarf shrub heaths" in which Cowberry and Bearberry are most typically represented is concentrated chiefly in the Cairngorms (granitic), the Monadhliath Mountains and along Speyside. Thus geological influence on plant distribution can be seen in maps having adequate detail and it is especially clear in ecologically exacting species and in the vegetation types they characterize.

Alternative habitats

Enough has been said to demonstrate that the ecological requirements of plants, which are satisfied by essentially local features of landscape such as minor topographic variation, soil drainage and irrigation, rock type, can be revealed in detailed maps of their distribution; and of course the larger the scale the clearer these influences become. For the sake of clear illustration examples have been given in which several species with near-identical distribution show coincidence with par-ticular soil conditions, but it would be false to imagine that ecologically similar species have the same total geographical distribution as a rule. When considering any partial area in detail, species can be expected to coincide if the same ecological factors govern them; but when their gross areas of distribution are compared, e.g. their occurrence over several states or across an entire continent, it can be seen that species fulfilling very similar ecological roles and occupying habitats of the same kind may belong to quite different geographical elements. For example, there are a number of plants which have a montane distribution like the calcicole species and grow together with them on cliffs and talus in the mountains but which also have a strongly coastal distribution at low elevations. In fact, as Mossy Saxifrage and Mountain Avens would be classed as montane species, Sea Pink (*Armeria maritima*), Sea Plantain (*Plantago maritima*) and Sea Campion (*Silene maritima*) would be recognised primarily as maritime plants from their distributions. None the less, they occur high in the mountains, in association with distinctly montane (arctic-alpine) species, showing that some montane habitats are as favourable as the coast in providing the particular ecological conditions they need and that the climate differences created by altitude do not inhibit them. Thus in British and Scandinavian mountain regions there is both ecological and geographical overlap between plants which elsewhere are associated with quite different situations.

We must be prepared to find that most plant species are capable of exploiting more than just one habitat and that some of the diverse environments they tolerate have features in common that are not immediately obvious. It is evident that coastal plants capable of growing in the mountains are not halophytes, i.e. they are not dependent on saline conditions, and even their widespread distribution in coastal habitats cannot be conditioned by this factor. Other maritime species demand saline conditions and for these the montane situations would be uninhabitable. Their alternative habitats are non-marine salt-flats located in continental interiors, e.g. some of the more saline "prairie-potholes" of South Dakota, and comparable saline lakes in Czechoslovakia (South Moravia). Thus comparison of distributions can bring out unsuspected differences among species that share the same habitat. The calcicoles on a mountain cliff of lime-bearing rock are there because in this situation their nutritional demands are satisfied but some associated species not notably calcicolous may be there because they cannot tolerate shading (e.g. by trees) and this requirement is also satisfied in the same habitat.

Scarcely any species occupies a particular habitat universally. In each habitat none of the possible occupiers have the same total territorial range with the result that on a transcontinental scale a whole series of communities can be seen as tenants, the changes between them being gradual as one species reaches its geographical boundary and gives place to another. The geographical distribution of species is more fundamental than that of vegetation types or animal communities which, after all, are the associations built up from the species available in each region. Knowledge of distribution and ecology are complementary. The full interpretation of a distribution map cannot be made without some knowledge of the ecology of the plant or animal, i.e. a description of its habits and behaviour and the situation where its conditions of life are met. Just as its form, shape and size are characteristics used to identify the species, so its geography or areal distribution is a unique attribute, presenting in an integrated way a summary of the influences of its environment and the effects of events in its history as a species.

SOURCES OF REFERENCE

Dony, J.G. (1967) *Flora of Hertfordshire.* Hitchin Museum.

Good, R. (1948) *A Geographical Handbook of the Dorset Flora.* Dorset Nat. Hist. Soc., Dorchester.

Horwood, A.R. (1933) *Flora of Leicestershire and Rutland.* Oxford University Press, London.

McVean, D.N. & Ratcliffe, D.A. (1962) Soils, Chapter 12 in *Plant Communities of the Scottish Highlands.* H.M.S.O. London.

McVean, D.N. & Ratcliffe, D.A. (1962) Plant geographical factors, Chapter 13 in *Plant Communities of the Scottish Highlands.* H.M.S.O. London.

CHAPTER 4

Climatic control of plant distribution

When examining the distribution of plants over territories larger than a single county, some of the effect of local conditions becomes obscure because it is impossible to show individual locations with accuracy when a smaller scale is used. However, a compensating feature is that factors which have a more general influence, irrespective of minor topographic and edaphic variations, become apparent although they cannot easily be detected within the limited areas of detailed survey. These factors are principally the influences of climate and of history. Since the latter is difficult to recognise, I shall discuss climatic factors first and deal with historical effects on distribution in later chapters.

Thermal influence on mountain plants

Many montane species, such as those mentioned in the discussion of edaphic influences, tend to occur at low altitudes in oceanic districts in the northern part of their range. Another peculiarity is that they frequently occupy the cliffs of north-facing corries (cirques) but not cliffs facing south even on the same mountain where the same rock strata outcrop on both sides of a ridge. Both these facts suggest a climatic factor, and observations by Dahl in Scandinavia indicate that temperatures above certain critical values may permanently damage many montane species. Dahl's attempts to substantiate this idea have become a classic example of biogeographic method. Although it has been studied with reference to mountain plants, the method and perhaps its findings may be relevant to the very similar problems of the distribution of high mountain fauna, especially invertebrate animals.

The microclimate as it affects plants on a rock-face is obviously very different in incident solar radiation on south and north exposures and even great differences in altitude cannot entirely compensate for aspect differences in high latitudes. Such differences are much less pronounced on mountains in low latitudes. There is no need to assume that limiting conditions are experienced continually or even frequently:

if the limiting temperatures occur on one or a few occasions annually this may be quite enough for them to be effective. Dahl therefore selected maximum summer temperatures, rather than monthly mean, as the relevant meteorological data. The coastal occurrence of mountain plants could be explained on this basis because the thermostabilizing effect of the sea ensures that even on exceptionally hot days the temperatures reached at the coast are never as high as those recorded even a few miles inland. The highest maxima in any district are therefore not felt in its coastal fringe.

As it is not practicable to obtain long-term records of maximum temperatures on the innumerable mountain cliffs occupied by these plants, the available standard meteorologic records must be used in the most appropriate way as substitute. Dahl compiled the mean maximum summer temperature for all Scandinavian stations over a fifteen year period because he believed it might be the highest temperature occasionally experienced that could irreversibly damage the leaves and growing points of these plants. He had in fact observed that the foliage of some species turns brown and dries out, even if well watered, when cultivated in botanical gardens. Though new leaves are produced the plant is forced to divert so much of its resources to the replacement of damaged leaves that it dies after a few summers at low elevations. As most meteorological stations are situated at lower elevations than where mountain plants grow it was necessary to correct the mean maximum temperatures by applying the lapse rate of 0.6°C reduction per 100 m of altitudinal ascent. This was done not only for the locations at which particular plants were known to grow but for all the highest points in the landscape, i.e. mountain summits and the hill-tops in lower country, using the data from each station to calculate the expected maximum temperature on all the summits in its vicinity. It was now possible on the derived data to construct isotherms which refer to expected mean maximum temperature at the highest elevations.

For example, the 25°C isotherm encloses that territory in which the hill-tops are likely to experience a summer maximum temperature of that value and less. Outside this isotherm no points in the landscape experience maxima as low as this: even the hill summits will be warmer than 25°C when extreme maximum temperature conditions occur.

The distribution maps for mountain plants can be arranged to form a series of "equiformal progressive areas". That is, starting with those restricted to the smallest area, the others can be superimposed so that each coincides with the one before and also overlaps its boundaries by a small margin all round. Dahl expressed this simply by saying that if you travel from southern or eastern Scandinavia towards the mountains you will encounter the alpine species in very nearly the same order, no matter what route you follow. If we compare the plant distributions and the isotherms for mean summer maximum temperature at the

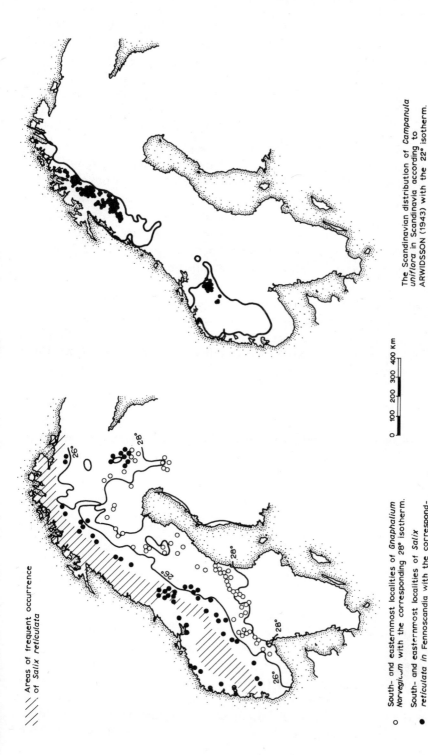

///// Areas of frequent occurrence
of *Salix reticulata*

○ South- and easternmost localities of *Gnaphalium
Norvegicum* with the corresponding 28° isotherm.

● South- and easternmost localities of *Salix
reticulata in Fennoscandia* with the corresponding 26° isotherm.

0 100 200 300 400 Km

The Scandinavian distribution of *Campanula
uniflora* in Scandinavia according to
ARWIDSSON (1943) with the 22° isotherm.

4.1   Correlations of plant distributions with mean *maximum* summer temperatures
in Scandanavia. (After Dahl, 1951.)

highest elevations, we find that their boundaries coincide in particular cases. Three examples are shown in Fig. 4.1. We have to remember that perfect coincidence cannot be expected because the plants actually experience a microclimate which should be measured near to the ground surface in the habitats where they live. A rock-face protected from direct insolation by its aspect can experience lower maximum temperature than the open summit of the mountain even hundreds of feet higher. It is therefore possible for a plant to exist in such places apparently outside the isotherm which defines its limiting temperature. This occurs in the case of *Salix reticulata* at a few localities (Fig. 4.1a). However, the great success of the method is that the vast majority of localities for any one of the montane species fall within the area described by one or another of these isotherms and often the distribution approximates closely to the boundary defined by maximum temperature (cf. *Campanula uniflora*, Fig. 4.1b).

Dahl (1951) was able to correlate the distribution of 150 species of montane plants with maximum isotherms ranging between 22°C and 29°C. As a rough guide these values are about 10°C higher than the corresponding July mean temperature. An extra virtue of this climatological approach is that the special isotherms have predictive value: each describes an area within which the temperature tolerance of a particular species is not exceeded at the highest elevations. The occurrence of the plant on every mountain within that area is subject to the existence of the appropriate habitat, especially considering its substrate and aspect requirements. Even when these requirements are fulfilled, whether or not the plant occurs at a particular location depends on whether or not it can reach that place by its normal means of dispersal.

Combined effects of summer and winter termperature

Another method of using climatological records to discover the nature of climatic factors limiting plant distribution was devised by Iversen (1944). Temperature records were collected for stations known to be approximately at the boundary of the distribution of Holly, Ivy and Mistletoe respectively, as well as for some stations both within and outside the present area of each species. The mean temperature of the warmest month was plotted (as ordinate) against the mean temperature of the coldest month (as abscissa) for each station to produce a scatter diagram. This climatic information was supplemented by direct observations of the plants themselves in the vicinity of all the stations included so that the temperature statistics could be related to their performance at these locations. If the plant could be found within 20 km distance of the station in flat country it was recorded and the

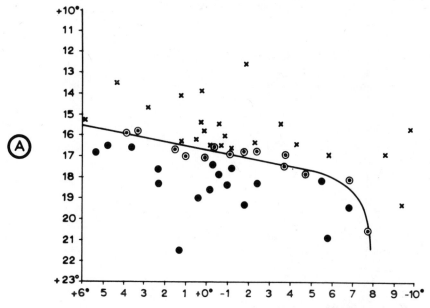

● *Viscum* within the station area ⊚ Station on the area boundary of *Viscum*

× *Viscum* absent from the station area

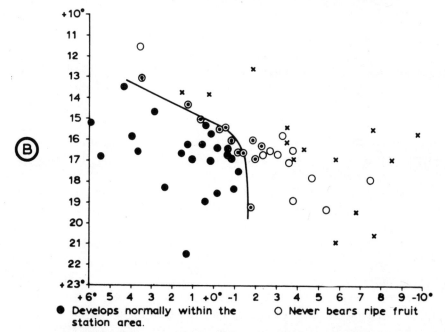

● Develops normally within the station area. ○ Never bears ripe fruit

⊚ Mostly sterile, only exceptionally bearing fruit. × *Hedera* missing within the station area.

4.2 Thermal limits at boundary of distribution in three European species (after Iversen, 1944). Mean temperature of warmest month (vertical axis) and mean temperature of coldest month (transverse axis) are plotted in relation to occurrence of (A) Mistletoe (*Viscum album*), (B) Ivy (*Hedera helix*), (C) Holly (*Ilex aquifolium*).

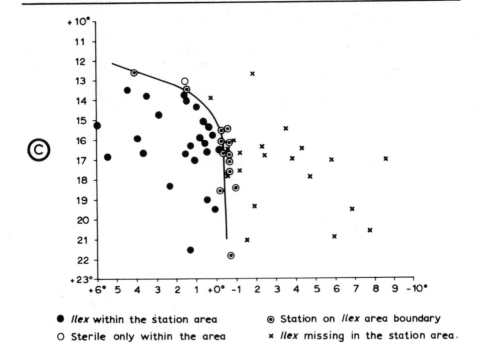

● *Ilex* within the station area   ◉ Station on *Ilex* area boundary
○ Sterile only within the area   × *Ilex* missing in the station area.

climate prevailing at that station was treated as applying also at the site where the plant occurred, providing that the difference in altitude between them did not exceed 40 m. Symbols are used on the scatter diagram to indicate the behaviour of the plant at each site, e.g. whether it was reproductive, present but infertile, or known to be at the boundary of the area for that species. On the diagram the symbols clearly separate those stations where plants were present from those where plants of the same species were absent, the stations for non-fruiting specimens or boundary sites falling within a narrow zone between. A line drawn through the points in this dividing zone defines the thermal conditions at the limit of the plants' distribution in terms of both summer and winter temperatures.

Mistletoe (*Viscum album*) is a semi-parasitic plant which grows high in the crowns of trees where it is fully exposed to air temperatures well above ground surface. Its curve (Fig. 4.2a) shows that it tolerates cold winter conditions (mean temperatures down to −8°C) in areas where summer mean monthly temperature reaches 20°C at least in the warmest month. Where summers are not so warm (e.g. 18°C) it is less tolerant of winter cold. The coolest summer conditions in which it can exist at all have mean July temperature above 16°C. This is reflected in its European distribution (Fig. 2.13).

Ivy (*Hedera helix*) is much less tolerant of cold winters and is never capable of producing viable seed where the coldest month has a mean temperature below −1°C. (Fig. 4.2b). Again in this species, as in the last, distribution does not solely depend on the severity of winter conditions: the warmth of the summer months is also a qualifying factor. Where no summer month has a mean temperature above 16° C, its tolerance of winter cold is even less. Ultimately in the mildest of temperate winter conditions (monthly mean 4.5°C), Ivy is limited by inadequate summer warmth.

Holly (*Ilex aquifolium*) has a thermal curve very similar to the last species (Fig. 4.2c). It is slightly less tolerant of winter cold and more tolerant of cool summers, being limited by the isotherm −0.5°C and surviving even where mean July temperature is only 13°C.

The graph method is a tool of great precision and makes it easier to interpret mapped distributions in relation to climatic factors. The Scandinavian distributions of Holly and Ivy (Fig. 4.3) demonstrate the geographical reality of the correlations deduced from the temperature curves. Both species are confined to a narrow coastal area by the generally north-south alignment of the winter isotherms (in this case isotherms 0° to −2°C are closely compressed by the rising ground of the mountainous hinterland). The distinct latitudes at which the two species are cut off northwards are determined by the summer temperatures of 13°C and 15°C July mean respectively.

I have referred to the question of whether a plant is fertile in discussing Ivy, and it is worth noting that sterility is characteristically the first sign that a species is near to its effective limit of distribution. The plant needs higher temperatures, or optimal temperatures prevailing for a longer period to produce ripe fruit and mature seed than it needs to sustain vegetative functions. Seed is dispersed from plants growing in satisfactory conditions to not far-distant locations where conditions are suboptimal. The seedlings may survive but these individuals can never produce seed themselves or only rarely in years of exceptional warmth. Thus at the limit of distribution there will often be a "twilight zone" in which the species exists but is not reproductive (see Chapter 8).

An example of a plant with a very clear northern limit in the south of England is Butcher's Broom (Fig. 2.2). Clement Reid (1899) said of this plant, "After watching its fruiting for twelve years in succession, I find that as a rule only one plant in fifty produces any fruit. These are few in number and as they ripen in November, an early onset of winter may prevent them ripening at all." The plant is perennial and hardy and can therefore survive but its successful reproduction may be very infrequent. The exceptionally warm summer of 1898 caused it to fruit freely in Hampshire and in that year Reid counted over forty ripe berries per plant. Its persistence in southern England under present conditions must depend on the intermittent occurrence of years warmer than average.

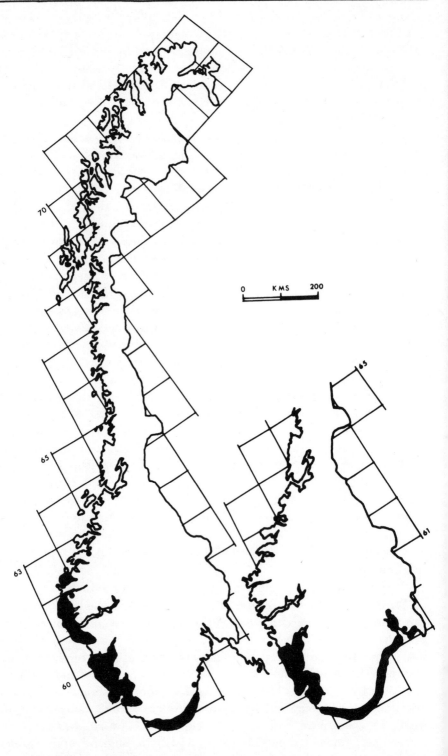

4.3   Distribution in Norway of (a) Holly (*Ilex aquifolium*), (b) Ivy (*Hedera helix*). After Faegri, 1960.

Enough examples have been mentioned to make it obvious that quite different aspects of climate may be effective in limiting the distribution of various plants even within the same territory. On reflection it is equally clear that different climatic factors operate restrictively in different parts of the total area of any one species. Thus while the eastern limit of Beech in Europe is controlled by the duration of low winter temperatures (a frost period exceeding four months), at its southern limit inadequate rainfall prevents its further expansion, 600 mm annual rainfall being necessary to meet its requirements. This explains why in the Cantabrian Mountains (northern Spain) it grows only above 500 m, in the Apennines (Italy) only above 1,000 m and in Greece only above 1,300 m. Sugar Maple in eastern North America is thought to be limited at its northern boundary by extreme cold (mean annual minimum −40°C) and by inadequate rainfall on the west (20″ isohyet in part). Some plants may need winter cold of at least modest severity (e.g. mean minimum values −10°C) for successful germination of seed. This is believed to control the southern boundary of Sugar Maple in the United States (Dansereau 1957).

Thermal influence on coast plants

The complex requirements of plants in regard to temperature can be detected in the distributions of many species providing that sufficiently detailed dot or grid maps are available. Plants of the coast occupy a number of well-defined habitats — strandline, dune, salt marsh, cliff and shingle. Each of these would appear to provide the same conditions wherever it occurs but the geography of coast plants discloses the fact that many are not ubiquitous. Even those familiar to each of us along coasts we know may reach a terminus only a few miles further on. What we had perhaps assumed to be representative of our entire coastline may prove to be unique in the light of biogeographical information.

The coasts of Northern Ireland and Galloway in south-west Scotland are a case in point. About a hundred miles of coast in both these regions contain a number of species never seen together elsewhere in the British Isles, for here some reach their farthest south and others their farthest north. Two species sharing the same rocky cliff habitats are Lovage (*Ligusticum scoticum*), here at its southern limit, and Rock Samphire (*Crithmum maritimum*) which occurs only exceptionally further north. Another species approaching its southern limit along the northern shores of the Irish Sea is the Oyster Plant (*Mertensia maritima*), always a rare plant of shingle. Along the same shores several species approach their northernmost limits: Cliff Spurrey (*Spergularia maritima*) on rocks, Sea Spurge and Portland Spurge on strandlines and dunes. The respective limits of all these plants on the east coast of

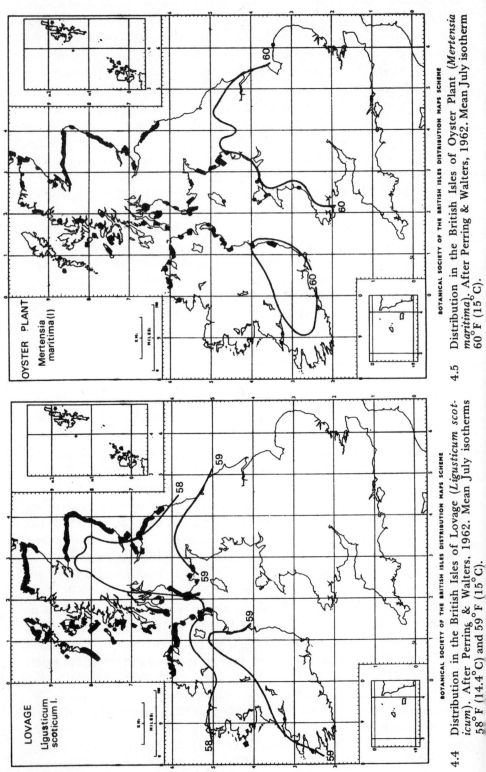

4.5 Distribution in the British Isles of Oyster Plant (*Mertensia maritima*). After Perring & Walters, 1962. Mean July isotherm 60°F (15°C).

4.4 Distribution in the British Isles of Lovage (*Ligusticum scoticum*). After Perring & Walters, 1962. Mean July isotherms 58°F (14.4°C) and 59°F (15°C).

Britain bordering the North Sea are entirely different. Only the long-term effects of climate can be responsible for controlling distributions in this way, especially since the climates of maritime environments are much more uniform than those of the interior due to the influence of the sea in moderating extremes of temperature.

The distribution of Lovage shows a simple relationship to isotherms for the mean temperature of the warmest month (Fig. 4.4). It is present only where mean July temperature is less than about 58°F. This implies that the plant cannot tolerate warmer summers than are usually experienced in these regions. Oyster Plant shows a comparable relationship to mean July temperature but in this case 60°F seems to be the limit of its tolerance (Fig. 4.5). The map includes records made before 1930 and the isotherms represent average temperatures for the period 1901-1930. As dated records indicate that this species has withdrawn northward since 1930 it would be interesting to make the comparison of its more recent distribution with the isotherms derived from temperature statistics for the period 1931-1960. In contrast high summer temperature seems to be necessary for the southern coastal plants but by itself this factor does not explain their distribution limits. Rock Samphire is absent from the eastern coast of Britain north of Suffolk but in contrast reaches the northern point of Ireland and occurs at isolated locations in the Hebrides and Sutherland (Fig. 4.6). This pattern with its strongly western bias reflects climatic gradient across the British Isles in winter. As the isotherm of mean January temperature 40°F almost completely demarcates this distribution we may conclude this plant is intolerant of colder conditions. Its needs for summer warmth are met everywhere in these islands, the mean July figure for Sutherland being only 55°F. The distribution of Sea Spurge illustrates more complex interactions of temperature, summer and winter (Fig. 4.7). On the east coast it extends north to the Wash and although it reaches further north on western coasts there is no simple relation with winter temperature. The east coast limit corresponds with the area of warmer summers (July mean >60°F): here the mean minimum temperature of the coldest month (February) is 34°F. On the west coast bordering the Irish Sea summers are not quite as warm (July 59°F) but winters are not as cold (February mean minimum 35°F). Further north this plant tolerates still cooler summers, ultimately less than 58°F mean July, but only in locations where the winter is correspondingly milder, i.e. with February mean minimum 36° − 37°F. This compensation between winter and summer temperatures reflects the length of the growing season. Ideally accumulated temperature should be measured in day-degrees above an arbitrary baseline and it is likely that a direct correlation could be achieved by this method. The lengthening of the annual period of growth westward is a feature of highly oceanic regions found on all continental margins in the latitudes of prevailing westerly air-streams.

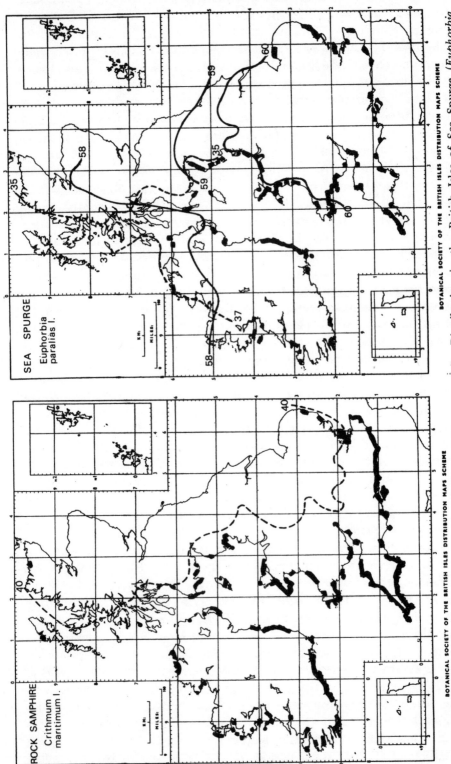

ROCK SAMPHIRE
Crithmum
maritimum l.

SEA SPURGE
Euphorbia
paralias l.

4.6 Distribution in the British Isles of Rock Samphire (*Crithmum maritimum*). After Perring & Walters, 1962. Mean January iso-

4.7 Distribution in the British Isles of Sea Spurge (*Euphorbia paralias*). After Perring & Walters, 1962. Mean July isotherms 58 – 60°F (14.4° – 15.5°C): continuous line. Mean *minimum*

Climatic correlation with vegetation

Geographers have often searched for correlations between climate and vegetation which might have universal validity. To achieve such correlations it is necessary to devise compound indices of climate and to treat only the most major physiognomic distinctions between vegetation types, for example, the boundary between forest and tundra (arctic timberline). As a rule of thumb Köppen considered that the northern tree limit in the arctic coincides with mean July temperature of 10°C (50°F). So far as information is available this appears to be a reliable guide to most arctic and alpine (i.e. altitudinal) timberlines in the north temperate and higher latitudes, but the reasons for its success may well be peculiar to this part of the world. It may not justify the assumption that other vegetation boundaries are capable of prediction from . climatic data. The holarctic realm .comprises the northern continents in extra-tropical latitudes and is bounded in the Old World by the Atlas-Himalayan mountain chains. Compared with other parts of the world its plant-life is very impoverished in number of species and in high latitudes this entire circumpolar belt is characterized by unequalled uniformity of vegetation. Consequently the species of trees that form the northernmost border of forests are very few in number and are all interrelated. Thus in North America from Alaska to Labrador the forest fringe is formed by an alder (*Alnus crispa*) which has its Eurasian counterpart in *Alnus incana*. In Scandinavia the forest limit is reached by a birch (*Betula tortuosa*). In fact a dozen or so species of willow, alder, aspen and birch can account for the northern timberline around the arctic, all of them members of two catkin-bearing families *Betulaceae* and *Salicaceae*. It is not surprising that they tolerate about the same growing season and extremes of cold. The alpine timberlines of the northern hemisphere outside the tropics are quite different in appearance and are under different climatic control from those of the arctic. The typical stunted, gnarled and dwarf stature of high mountain timberlines (krummholz) displays the limiting influence of wind on growth. The species differ altogether from those of the arctic, but again show relationship between themselves in that most are species of pine, e.g. Limber Pine (*Pinus flexilis*) in Utah, Mountain Pine (*Pinus mugo*) in the Alps and Carpathians, or fir, e.g. Fraser Fir (*Abies fraseri*) in the Appalachians, Silver Fir (*Abies pindrow*) in the Himalayas. The temperature regimes of these elevated regions bear little resemblance either in diurnal or annual cycle to the conditions of high latitudes, so that if the 10°C July mean does approximate to alpine timberline the operative factors can hardly be the same as those which coincide with that isotherm in the far north.

In the tropics the tree-limit is attained at high elevations on the peaks and plateaux of many mountain ranges and on isolated summits (e.g. of volcanic origin) and it is therefore often an interrupted and localised phenomenon. The tree species that form the forest edge differ from one region to the next. They often include species that are endemic* within a mountain group and these do not show generic or familial relationship between the timberline trees of different continents. Indeed, even the definition of timberline on tropical mountains can be a problem because of the peculiar life-forms encountered there and variations in the spacing and height of individuals. These relate more to wetness or dryness of climate than to temperature. The difficulty of defining arctic timberline is illustrated in Fig. 1.5.

Of course, there may be practical reasons for establishing empirical relationships between vegetation and climate which hold true only within specified regions. For example, Holdridge has calculated a climatic parameter which he terms biotemperature by dividing by twelve the sum of all positive mean monthly temperatures measured in degrees Celsius. Those months at any station which have a mean temperature below zero are not counted into the total. Working in the American tropics he equated the altitudinal timberline with the biotemperature of 3°C. In Alaska, on the other hand, Thompson found that this figure could not be applied to the montane timberline between 63° and 56°N latitude. Instead a modification was adopted, again empirically, in which only mean monthly temperatures above 10°C were totalled and then divided by twelve. On the basis of this calculation the *zero* biotemperature corresponded well with the upper limit of forest in *that* region. By recognising the practical usefulness of such correlations for special purposes we do not sustain any misconceptions about their cause or their universality. Consequently we need not endorse the outdated view that vegetation everywhere, regardless of its composition, responds in a standard way to the influences imposed by climate, as though according to some natural rules. Such assertions are due to subjective judgements in deciding which vegetation boundaries in different continents are the corresponding representatives of "timberline" or "desert" etc. The crude approximation of vegetation types with climatic parameters on small-scale maps is no guide to the real relations of climate and vegetation.

SOURCES OF REFERENCE

Dahl, E. (1951) On the relation between summer temperature and the distribution of alpine vascular plants . . . *Oikos*, vol. 3, part 1, pp. 22-52.

Faegri, K. (1960) Maps of the distribution of Norwegian Vascular Plants. Volume 1, Coast Plants. Oslo.

Holdridge, L.R. (1964) Life Zone Ecology. San José, Costa Rica. (Cited by Thompson).

* Endemic means exclusive to or not occurring naturally elsewhere.

Iversen, J. (1944) Viscum, Hedera and Ilex as climatic indicators. *Geol. Fören. Stockh. Förh* 66, 463.

Reid, Clement (1899) *The Origin of the British Flora.* Dulau & Co., London.

Thompson, W.F. (1969) A test of the Holdridge system at sub-arctic timberlines. *Proc. Assoc. Amer. Geog.* 1, 149-153.

CHAPTER 5

Vegetation: methods of description

As vegetation is one of the natural attributes of any place (as are rock, soil, water, weather), the distribution of different kinds of vegetation over the earth's land surfaces is one of the fundamental geographical phenomena and forms one aspect of biogeography. It differs from other aspects of distribution already considered because it is compound in character and because it requires area for its expression. In preceding chapters the subject of distributional study has been the species, a unitary concept comprising all the individuals recognised as being of the same kind. In vegetation we are confronted by heterogeneity: at any given place it is composed of a variety of individuals belonging to different species and yet as a group possessing certain characteristics that make it recognizable from place to place wherever it occurs. The distribution of vegetation is the geography of plants considered in their natural associations.

Plant size, spacing and numbers

Because the unit of vegetation is a group and not an individual it requires a certain area for its expression and for recognition. If only a few species are represented in the community that area may be small; probably 10 square metres will adequately represent the community of unforested bog vegetation that occurs in highly oceanic districts in western Europe, in which perhaps only twenty plant species are present. In a community with larger numbers of constituent species the minimal area necessary for it to be adequately represented must be larger. For example, in remnants of the original prairie of Kansas the number of species found was 40 in a study area of 300 sq ft. Fifty square metres might be considered representative in vegetation of this type. It is difficult to obtain comparable information but the contrast between these vegetation types is illustrated by Weaver's statement that one square mile of prairie contains about 200 species and a report by Osvald* that only 40 species can be listed from an equivalent area of

---

* He recorded 40 species of vascular plants; also 20 bryophyte species and 13 lichen species in one square mile of Komosse.

bog. Further allowance must be made for the stature of plants involved and the spacing of individuals. Increase in either size of plants or distance between plants' necessitates still greater areas for the community to be adequately represented. Comparison of forest with prairie illustrates the effect of greater plant size. In forests of Europe and temperate North America, where species numbers are not excessive, plots of 500-1000 sq m are representative. An estimate from an account by Tansley suggests a total of 80-100 species of all kinds, including bryophytes, for a plot area of 10,000 sq ft (929 sq m) in English oakwood. In the tropical forests of central America and south-east Asia the sample area must again be enlarged, probably by a factor of ten, to represent adequately the far greater number of species. On plots of 1 hectare (10,000 sq m) between 150-200 different tree species alone are present in many parts of the Malaysian forest and these figures exclude consideration of all plants with stems less than 10 cm diameter, for practical reasons! The effect of plant spacing in this matter is illustrated by desert vegetation or other types in which there is an "open" community and plant canopy provides incomplete cover of the ground surface. The saltbush vegetation of arid plains (playa) in New Mexico yield only 30 plus species in an aggregate area of 1,000 sq ft (93 sq m) from dispersed subsamples (cf. prairie).

This discussion has introduced several of the distinctive features of vegetation that may be used as criteria for describing and recognising various types, namely, the list of constituent species (whether many or few), the stature — size and shape — of the included plants, and the spacing of individuals relative to one another in both areal and vertical dimensions (structure).

Floristic description

While a complete or at least representative list of the species present is a sure means of characterizing any vegetation, for practical purposes this must be abbreviated to just a few names for ease of reference, and obviously the species chosen should be those of most diagnostic significance. This is the approach of plant sociology or phytosociology, originated in France and Sweden by Braun-Blanquet and Du Rietz respectively and now generally used by botanical ecologists in continental Europe. Whatever its merits, it does not lend itself to application by non-specialists because diagnostic species are not in all cases the most conspicuous. Although not capable of distinguishing vegetation differences on a fine scale, a system based upon naming the dominant species or principal dominants, where there are several, can be applied without specialised botanical knowledge and is sufficient for most geographical description at regional scale. For example, we denote the original forest of the Great Lakes region as the Maple-Beech-Hemlock association.

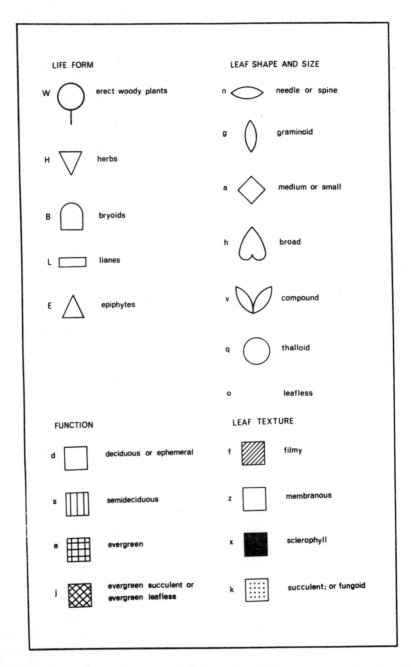

5.1 Key to the life-form symbols devised by Dansereau. (After Dansereau & Arros, 1959.)

Life-form description

An alternative means of describing vegetation is to examine the forms of the plants that comprise it. First, however, a scheme must be prepared to describe and name the types of plants to be recognised and also those particular features which appear to be important in characterizing the vegetation. Several classifications exist, some of them with accompanying symbols so that the vegetation may also be represented in diagrammatic form.

In the system devised by Pierre Dansereau the plant is first assigned to one of five primary categories on the basis of its general form. Each category can be denoted by a code letter in addition to the graphic symbol used in diagrammatic vegetation profiles (Fig. 5.1).

The five primary categories are:

W   erect woody plants (trees and shrubs)
H   herbs (non-woody plants) both broad-leaved and graminoid
L   lianes (climbing plants) and others that do not support themselves
E   epiphytes
B   bryoids

The fact that the use of these terms does not correspond to the orthodox definitions points to the difficulty of finding any half-dozen categories which can satisfactorily embrace the multitudinous variety of plant forms. Dansereau has himself abandoned a distinction that was part of this scheme for many years, i.e. separate categories and distinctive symbols for trees and shrubs. In their place we now have only a single category (W) to include all erect i.e. mechanically self-supporting, woody plants. This is undoubtedly an improvement, for the borderline between a tree and a shrub is hard to define in strict terms. For instance, does height play any part in the distinction or does a shrub always have multiple stems? With herbs (H) are included grasses and grass-like plants (sedges, rushes, etc). Special interpretation of the term lianes (L) is made by Dansereau and it does not correspond to the strict definition i.e. woody-stemmed climbing plants such as Ivy (*Hedera*). In category (L) he includes twining plants whose stems are herbaceous (e.g. Bindweed, *Polygonum convolvulus*), others which are merely scrambling – sometimes with the aid of hooks or spines (e.g. *Strychnos)* and some which adopt a comparable habit although technically they are parasites (e.g. Dodder, *Cuscuta* spp.). These are in addition to the vines or true lianes.

The term epiphytes (E) is similarly used of plants not included in the classic definition, i.e. not rooted in the soil but attaching themselves to the aerial parts of other plants. Dansereau includes here plants which have this characteristic but are also parasitic upon their hosts (e.g.

Mistletoe, *Viscum, Loranthus* spp.). Category (B), "bryoids", is used for all plants possessing moss-like form and thus includes certain flowering plants of "cushion" habit, (*Silene acaulis, Raoulia australis*) as well as true mosses (many of which have a spreading form). Lichens and other closely encrusting growths on rock surfaces (epilithic) are treated with epiphytes. Hence the triangular symbol is used in the diagram for tundra (Fig. 7.1). It becomes hard to see just where the limit is set to the inclusion of transitional forms. These tend to blur the edges of the categories, however distinct they may appear at first sight.

Following this primary classification, the height above ground which the plant reaches is recorded in one of seven classes and is illustrated conventionally by the height of each symbol in the diagram (Fig. 5.2).

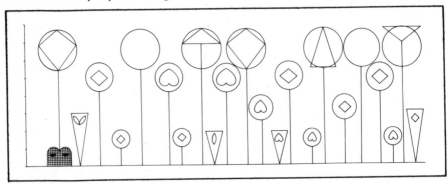

5.2   Diagrammatic structural profile illustrating the use of Dansereau life-form symbols and height classes. Modifications of the basic symbol to illustrate distinctive forms of tree crown are shown in the highest stratum, (after Dansereau, 1958).

7.	more than 25 m
6.	10 – 25 m
5.	8 – 10 m
4.	2 – 8 m
3.	0.5 – 2m
2.	0.1 – 0.5m
1.	0 – 0.1m

The merits of this particular height classification become clearer in relation to vegetation profiles rather than to individual plants, for which a simpler scheme with more regular progression of the intervals might seem more rational. Since foliage characteristics, together with height and stature, are of paramount importance in determining the general physiognomy of vegetation, Dansereau next considers this in several aspects. As "function" we select from four alternatives which describe the period or circumstance in which the plant remains green, i.e. is functioning vegetatively. These are (i) deciduous or ephemerally green – bearing foliage only seasonally or for a shorter period; (ii) semi-deciduous; (iii) evergreen; (iv) evergreen-succulent. The last includes certain forms which never bear leaves but which remain perennially functional because photosynthesis is carried out in stem tissues.

There are six categories to describe leaf form and size (Fig. 5.1) which are self-explanatory except perhaps for "thalloid" The term applies to liverworts and lichens, in which leaves are not produced, and presumably also to certain flowering plants of peculiar form in which the normal distinction between stem and leaf is obscure and a flattened plant body (thallus) of irregular shape results, generally possessing a succulent texture, e.g. some *Mesembryanthemum* species. With so few categories to accommodate all the forms of foliage encountered in plant life throughout the world it is apparent that these terms must be applied rather loosely in some cases, a not entirely satisfactory feature of the scheme. Illustrating this point, the needle leaf type describes the foliage of many conifers but there are possibly just as many whose leaves cannot be described as needle-like yet none of Dansereau's other categories are appropriate. Those of cypress trees (*Cupressus*) are small but thick, scale-like and closely overlapping. Similar leaves are common in other conifer families and are called "cupressoid". There are others which might be difficult to describe in Dansereau's terms, such as the foliage of Monkey-puzzle tree (*Araucaria araucana*), native to Chile but introduced as an ornamental tree and widely planted in parks and gardens in western Europe.

Finally leaf texture is described in four alternative terms. Filmy texture describes the most delicate type of leaf which is thin and translucent as in the filmy ferns (*Hymenophyllaceae*). Papery (or membranous) describes the texture of most broad-leaved deciduous foliage, which is not thickened, leathery or rigid. Leathery (or sclerophyllous) applies to leaves that are somewhat thickened and rather stiff or even hard and rigid. It therefore includes for example most needle leaves and also the broad leaves of many chapparal species. Succulent describes leaves that are thick and fleshy in texture and juicy when crushed, as in many stonecrops (*Sedum* spp.). In practice this term is not only applied to plant foliage since stems and branches can have this character too. Thus a cactus or cactus-dominated vegetation would be described as evergreen (e), leafless (o), succulent (k), according to Dansereau.

The scheme apparently employs only simple distinctions between plants which are easily observed. However, considering all of the diverse situations, types of vegetation and plant form to which it must be applied, in practice plants of conspicuously varied form will have to be assigned to the same descriptive category. For example, herbs with deciduous graminoid leaves (Hdg) must include grasses of the temperate zone and also presumably those large monocotyledonous herbs with sword-like leaves such as irises (*Iris*) and cat-tail (*Typha*). This descriptive scheme therefore cannot always convey a visual impression of any particular vegetation: the impression gained will be a generalized one.

The limitations of essentially simple life-form classifications are by now apparent. A more flexible scheme is that of Emil Schmid (1957). It may be used to describe vegetation in whatever degree of detail is required without any sacrifice of accuracy in depicting each species included.

In essence Schmid's scheme provides a do-it-yourself kit for the description of plant growth-form and is particularly designed for diagrammatic presentation using conventional symbols (Fig. 5.3). No arbitrary distinctions are made — separate categories for tree and shrub are not used — but thorough observation of more particular features is demanded. We start with the above-ground parts of the plant and analyse the branch system. A primary axis is recognised, e.g. the bole of

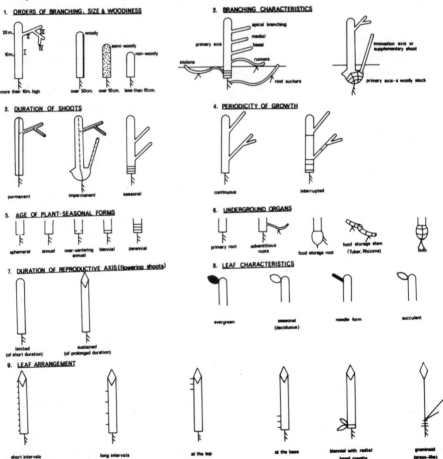

5.3 Key to the life-form symbols devised by Schmid (redrawn from Schmid, 1957.)

a tree, or the erect shoot of a flowering herb, and then successive orders of branching are counted excluding short shoots and twigs less than

20cm in length. In this way differences in the characteristic arrangement of shoot systems can be recognised between species which in
other respects appear to be of the same form. Height-classes are
distinguished as usual and the relative position of branching on the
plant can be shown conventionally as basal, medial, apical or any
combination of these. The branch system or different parts of it can be
shown as woody (fully lignified), non-woody (not lignified) or as
semi-woody (partly lignified) as in the trailing shoots of brambles
(*Rubus* species). On this scheme no formal distinction between shrubs
and trees is necessary and plant forms such as vines (lianes) and
brambles can be accurately represented. It would be difficult to assign
these to any category in a simpler classification. The arrangement and
type of foliage is also considered in some detail, showing for example
whether leaves are arranged on the shoots in radial symmetry, as in ash
trees (*Fraxinus*) and oaks (*Quercus*) or in dorsiventral symmetry, i.e.
the shoot as a whole is flattened with an obvious upper and lower side,
as in beeches (*Fagus*). Other refinements, indicated in the key
(Fig. 5.3), provide an armoury of technique for the clear description of
vegetation in terms of its physiognomy.

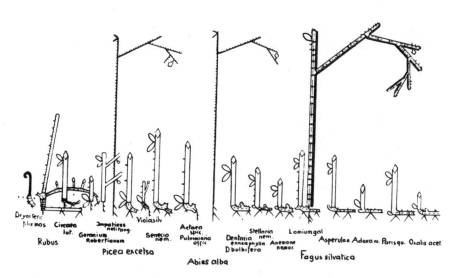

5.4 Example of a structural profile for Beech forest illustrating the use of
Schmid's life-form system. (Reproduced by permission)

Functional aspects of life-form

A feature that distinguishes Schmid's treatment of growth-form is the
importance attached to the behaviour or functioning of the plant in

regard to time. The longevity of the whole plant is shown e.g. annual, biennial, perennial, and the duration of various parts of the shoot system, and whether growth is continuous (as in the ever-wet forest of the tropics) or interrupted (by an unfavourable season). Underground storage and perennating organs, and specialised shoots (e.g. runners, stolons and others) which are effective in vegetative spread are also dealt with in this comprehensive scheme. By this kind of description (Fig. 5.4) Schmid has emphasized the essential importance of function and periodicity in plants to the vegetation which they collectively create. It must remind us that the static description of external anatomy is scarcely enough for an understanding of vegetation and with regard to adaptation may conceivably lead to false conclusions.

There is another kind of life-form system that differs fundamentally from both Dansereau's and Schmid's schemes. Whereas these rely upon the combination of many independent characteristics, the system devised by Carl Raunkiaer relies upon the possible variations in just one characteristic. To discover a single attribute of plant form and function which is so fundamental that it is relevant to all plants in whatever region and situation is a feat of imaginative thinking. Carl Raunkiaer first published his ideas on this subject in 1903 and 1907 in Danish, and although reviewed briefly in English (1913) they did not become available in full translation until 1934.

Raunkiaer set himself the objective of discovering in plant form some feature that would betray the plant's adaptation or fitness to survive in the particular climate which it experiences. Functional attributes are therefore intrinsic to Raunkiaer's descriptive scheme. Seeing that the most critical aspect of climate affecting plant growth was to be found in seasonal extremes of temperature and aridity, he decided that the most susceptible part of the plant would be the special growing tissues in which all new cells are produced. These meristematic tissues are located in the buds from which all new leaves and extension of shoots originate. He devised a system of classifying plants according to the degree of protection afforded to the perennating buds. He believed that the height above ground at which these buds survived the unfavourable season was an inverse measure of their protection because it "has a very definite bearing on their relationship to humidity". We shall discuss this interpretation further when the details of Raunkiaer's classification have been given. Whether its value as an index of adaptation to climate is accepted or not, the fact remains that Raunkiaer provided us with a descriptive tool of universal application in the plant world. The principal categories of life-form named by him are as follows.

Phanerophytes (P) are plants whose branch systems project freely into the air, bearing the perennating buds relatively exposed to prevailing meteorological conditions at various heights above ground surface.

They are sub-divided into height classes, thus:

Mega-phanerophytes	(Pg)	more than 25 m
Meso-phanerophytes	(Pm)	10 − 25 m
Micro-phanerophytes	(Pp)	2 − 10 m
Nano-phanerophytes	(Pn)	0.5 − 2 m

There is a special category of Climbing Phanerophytes (Ps) which bear their buds at some height above ground by virtue of their climbing habit, i.e. supported by other phanerophytes.

Notice that this classification makes it unnecessary to distinguish between "tree" and "shrub" and between woody and non-woody stems. Qualifying description takes account of whether the buds are naked, i.e. primordial leaves are exposed, or protected by a covering of bud scales, as are the buds of most temperate north hemisphere plants.

Chamaephytes (Ch) are plants which carry their perennating buds in the air at a lower level than phanerophytes, namely 0-0.5 m above ground surface. They include both woody and herbaceous perennials, in a variety of forms from dwarf shrubs to cushion plants.

Hemi-cryptophytes (H) have the surviving buds situated actually in the surface layer of the soil, i.e. protected by the soil or by leaf litter. The aerial shoots that carry foliage and inflorescence survive only for a single growing season, then die back to soil level. Such shoots are normally herbaceous (non-woody) and most biennial and perennial herbs belong to this category. Subdivisions based on growth habit can be made, for example, caespitose (Hc) with tufted erect leaves growing from a root stock at soil level, and rosette (Hr) with leaves arranged in that form and adpressed to the ground surface. These can give a useful impression of the plants' appearance during the growing season.

*Geophytes** (G) include plants whose surviving buds are always beneath the soil surface and the perennating organ is some form of underground stem or rootstock, e.g. bulbs, rhizomes, tubers. These types are protected from atmospheric conditions during the resting season.

Therophytes (Th) are plants which survive the unfavourable period as seeds. In this form the plant is least vulnerable to climatic extremes. This category includes all plants in which the entire plant body is short-lived, growing from seed to reproductive maturity within the span of a single favourable season. These are annuals and also ephemerals, in which the life-cycle is shortened to such an extent that several generations may be completed within a single "growing season" of temperate climates. In desert climates favourable conditions of much shorter duration can suffice for the growth of at least one generation.

Hydrophytes (HH) were considered by Raunkiaer as equivalent to geophytes but have their perennating buds in the mud beneath water,

* Also known as cryptophytes (Cr).

which provides a comparable degree of protection. The only additional special categories necessary to complete this classification are epiphytes (E), which exist high above the ground on the branches and trunks of phanerophytes, and stem succulents (SS) which Raunkiaer regarded as a peculiar type within the phanerophytes.

Leaf-size statistics

Among features of plant form that can be used to characterize distinctive types of vegetation, leaf size must be mentioned. Raunkiaer devised a classification which is still in use and, in fact, attracting renewed interest in modern descriptions of vegetation. He proposed a series of six classes of leaf surface area increasing on a logarithmic scale, but in a revision by Webb (1959) a seventh class was introduced by splitting the large size interval originally included in "mesophyll".
Therefore, we now have:

Leptophylls	up to 25 sq mm
Nanophylls	up to 9 x 25 sq mm (= 225 sq mm)
Microphylls	up to 9^2 x 25 sq mm (= 2025 sq mm)
Notophylls	between 2025 and 4500 sq mm
Mesophylls	up to 9^3 x 25 sq mm (= 18,225 sq mm)
Macrophylls	up to 9^4 x 25 sq mm (= 164,025 sq mm)
Megaphylls	greater than 164,025 sq mm

In survey the leaf size class to which every species belongs is determined by measurement, the individual leaflet of compound leaf types being regarded as the unit equivalent to a single leaf of simple type. A rapid means of doing this is to measure both length and breadth of leaf blade and to compute areas as $\frac{2}{3}$ lamina length multiplied by breadth (Grubb *et al*, 1963). The number of species falling within each size category is compiled and in comparisons between vegetation types interesting differences emerge.

In the lowland tropical forest of Ecuador the mesophyll category is best represented and a generally similar spectrum (proportionate representation) applies in lower montane forest there. However, in the montane forests while mesophylls remain predominant the number of notophyll and microphyll species increases and the number of macrophylls declines. In upper montane forest this tendency develops further: a majority of species belong to the microphyll class and species with leaves larger than notophyll may be absent altogether (Table 2). Thus in tropical regions larger leaf sizes characterize low altitude forests and smaller leaf sizes those at high altitude. The importance of compound leaves also changes with altitude, these accounting for 36% of species at low elevations and only 9% at high elevations in Ecuador (Grubb, loc. cit.)

Table 2 Leaf-size spectra compared for various tropical forests (data from Grubb and Webb)

	Micro-	Noto-	Meso-	Macrophyll
Lowland forest, N.E. Australia	2	39	59	—
Lowland forest, (Mucambo) Belem, Brazil	16	31	53	—
Lowland forest, (Shinguipino) Ecuador	9	14	50	27
Montane forest, (Borja) Ecuador	13	26	57	4
Upper Montane forest, N.E. Australia	95	5	—	—

These examples demonstrate that leaf size statistics are useful in expressing the character of vegetation numerically from survey of representative samples. It is of course equally applicable to non-forest vegetation.

Applications to vegetation survey

In the course of survey the life-form of each species present must be described. It has been claimed as an advantage for physiognomic methods that the observer need not know the names of the plants he encounters. On the other hand a non-botanical observer would probably recognise fewer kinds of plant than a botanist because the finer distinctions which are diagnostic for some species would escape notice by the non-specialist. Since attention is concentrated on the vegetative features of the plant in life-form survey, plants belonging to unrelated species may be confused because of similar forms and foliage unless they are examined in greater detail than is needed simply to denote their life-form. The result could be that only one kind would be recorded where in reality several species existed.

It may be asked whether such variations really matter when the object is to characterize the vegetation as a whole of which the species and life-forms are merely component parts. To some extent the answer depends upon how the information on life-form is to be used. When a life-form survey has been completed we have an inventory of the plant constituents but this may be in one of two forms. In one case the vegetation has been searched with the aim of recognising the botanical species present, without regard for the quantities in which they occur, and their individual life-form characteristics have been recorded. The results are analysed in terms of how many species are present in each of the life-form combinations described and, of course, they may be expressed in relative proportions on a percentage basis. Treated in this way no special importance is given to the form of plant which is most abundant on the ground, i.e. represented by the greatest number of individuals, or to the plant which may structurally dominate the vegetation. However, failure to distinguish all species present unavoidably distorts the proportionate representation of forms. It may therefore be argued this approach is for the botanist!

In the second method the quantitative importance of each species or form-unit in the vegetation is measured in the course of survey, usually by estimating its percentage cover of the ground. This may be represented conventionally in a structural diagram by the use of symbols in appropriate proportions. Many types of vegetation are markedly stratified and in the absence of a structural diagram (on which height-classes are shown) the life-form composition of each layer should be described separately, otherwise the implication of percentage cover values becomes obscure. For example, in West African savanna there is an interrupted or discontinuous layer of deciduous thick-barked trees standing above a field layer of tall grasses. In order to report accurately the low density of trees their percentage cover must be recorded separately from cover values for plants of similar life-form in lower strata. If Raunkiaer's life-forms are used then plant height is recognised intrinsically, e.g. in the subdivision of phanerophytes.

Adaptation to the prevailing environment can be claimed for the life-form exhibited by the largest number of species, e.g. in savanna or forest etc. but also for the life-form of those species which dominate the vegetation structurally or numerically. This question aside, both methods of survey provide the means of describing and comparing vegetation from place to place. A word of caution is necessary, however, as the emphasis of the second method outlined is obviously placed on the structural and numerical dominants and tends to neglect the contribution of species which are individually rare but which collectively may be important. It also risks neglecting seasonal aspects of vegetation e.g. plants which remain dormant and invisible in the soil except during a limited period of vegetative activity. The garrigue vegetation of the Mediterranean region consists predominantly of low-growing leather-leaved shrubs (*Pistacia lentiscus, Daphne oleaefolia*) and aromatic herbs but in the period February—April its appearance is transformed by the sudden emergence of shoots and profuse flowering of many species of geophytes, i.e. plants persisting from one growing season to another through intervening periods of drought only underground (as bulbs, tubers, and rhizomes). The time to see the floristic richness of such vegetation, and to assess it in terms of number of species and variety of life-form is therefore early in the year and this would be important in carrying out any kind of census.

All of the schemes described can be applied to a *taxonomic* analysis of vegetation and with equal validity to a *physiognomic* analysis. In the first we try to answer the question, "How many kinds of plants here are phanerophytes, or microphylls, or evergreens . . .?" In the second case, the question is "How abundant are phanerophytes, etc. in this vegetation and what role do they play in its structure?" In making use of such classifications as tools for the clear description of vegetation we need not feel compelled to apply them only in the way chosen by their original authors.

SOURCES OF REFERENCE

Grubb, P.J., Lloyd, J.R., Pennington, T.D. & Whitmore, T.C. (1963) A comparison of montane and lowland rain forest in Ecuador. I. The forest structure, physiognomy and floristics. *J. Ecol.* 51, 567-601.

Osvald, H. (1923) Die Vegetation des Hochmoores Komosse. *Svenska Vaxtsoc. Sallsk. Handl.* I. *Akad. Abhandl.* Uppsala.

Potter, L.D. (1957) Phytosociological study of the San Augustin plains, New Mexico. *Ecological Monographs 27*, pp. 113-36.

Raunkiaer, C. (1907) The life-forms of plants and their bearing on geography, pp. 2-104 in his collected essays published by Clarendon Press, Oxford (1934).

Raunkiaer, C. (1916) The use of leaf size in biological plant geography, pp. 368-78 in his collected essays. Oxford 1934.

Schmid, E. (1957) Ein Vergleich der Wuchsformen im Illyrischen Buchen- und Laubmischwald. *Bericht d. Geobot. Forschungsinstitut Rübel* 1956. pp. 66-75 Zürich.

Tansley, A.G. (1949) *The British Islands and their Vegetation.* Cambridge, Cambridge University Press.

Tomanek, G.W. and Albertson, F.W. (1957). Vegetation on two prairies in western Kansas, *Ecological Monographs 27*, pp. 267-81.

Weaver, J.E. (1968) *Prairie Plants and their Environment.* University of Nebraska Press, Lincoln.

Webb, L.J. (1959) A physiognomic classification of Australian rain forests. *J. Ecol.* 47, 551-70.

CHAPTER 6

Vegetation: the question of adaptation

As a sequel to an account of plant form *per se,* we must discuss the question of adaptation to particular environmental conditions, since geography is largely concerned with the relationships between observed phenomena and the processes of their environment. Vegetation geography is still encumbered with ideas on this topic which were current in botanical science half a century ago and which at that time represented an attempt to apply the knowledge gained from early experimental work in plant physiology to "explaining" the still earlier descriptive accounts of plant anatomy and morphology. We need to recognise the contradictions that exist in the distribution of plant forms and, having done this, to reappraise our interpretation.

The needle leaf

The needle leaf is so familiar in northern hemisphere conifers — spruces, firs, larches and pines, for example — that it becomes mentally associated with the climates of those regions where forests of such trees predominate in the natural vegetation. With no more justification than this, there is a tendency to assume that this form of foliage is adapted above all to existence in the harsh climate of the boreal zone, especially in relation to its short growing season and its long and severe winter cold. Indeed it is not difficult to invent reasons which appear to explain how the needle leaf is suited to these conditions but to believe that adaptation is therefore involved is pure teleology. A more rational approach is to consider where, throughout the world, needle leaves are a prominent feature of the vegetation and what groups of plants possess them. Certainly they are constant characters in some families of conifers, notably the pine family and the yew family (Pinaceae and Taxaceae), which are widely distributed in the northern hemisphere, but which also have member species in sub-tropical and even tropical regions. Other conifer families, while possessing their share of needle-leaved species, for example junipers, also display other forms of leaf

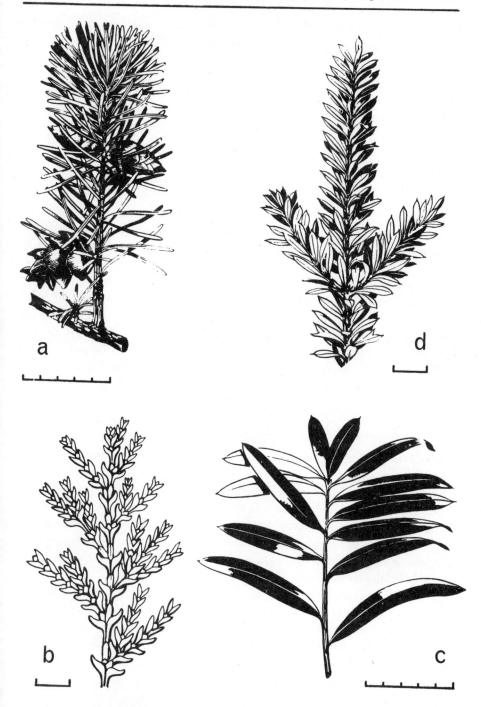

6.1 Variety of leaf form among conifers. (a) needle leaf characteristic of the pine family, (b) overlapping scale-like leaf ("cupressoid") illustrated by *Austrocedrus chilensis;* broad-leaf types illustrated by (c) *Podocarpus elatus* and (d) *P. nivalis.* Drawings by R.H.W. Herbert, by permission. Each scale division represent 1 cm.

such as the "cupressoid" type and even flat broad-leaved evergreen foliage, especially in the southern hemisphere. The variety of leaf form among conifers is illustrated in Fig. 6.1.

The greatest variety and concentration of coniferous trees in terms of the number of divergent kinds (families and genera) occurs in middle and low latitudes, especially in south-east Asia and Australasia including the island groups of Melanesia. Here we find conifers belonging to *Podocarpus, Agathis, Araucaria, Dacrydium, Phyllocladus, Libocedrus, Callitris* and fifteen less widespread genera in contrast with the few represented in high latitudes (*Abies, Picea, Larix, Pinus*) and this suggests that the group as a whole evolved and differentiated within the tropics. The needle leaf form is well represented among the conifers of low latitudes in the Old World (Fig. 6.2) and it would seem that this characteristic also evolved here, i.e. in a region where seasonal change in climate is minimal and extremes of drought and cold do not occur. In south China and Formosa within the sub-tropics and from the Philippines to New Caledonia within the tropics, the relatives of Yew are clearly recognisable by the close resemblance of their needle leaves to those of northern species. Unrelated trees of other coniferous families which also possess needle leaves in moist warm climates are, for example, *Cunninghamia lanceolata, Podocarpus totara,* and the so-called Norfolk Island "pine" (*Araucaria excelsa*) which is not in fact a pine strictly speaking. In the boreal forests of the northern hemisphere there are merely half a dozen conifer species respectively in North America and in Eurasia which because of their vast numbers of individuals and the correspondingly large area they occupy attract an exaggerated importance to the needle-leaved trees of high latitudes. Indeed, we might well express surprise that so few species of this form have found themselves equipped to exploit these northernmost forest lands. It is the firs (*Abies*), spruces (*Picea*) and larches (*Larix*) which dominate the northern forests as they do also the forests of high elevations in the great mountain ranges of middle latitudes (Rockies, Appalachians, Himalayas).

Pines of many species occupy territories in warm temperate and sub-tropical latitudes: in fact the greatest number of *Pinus* species (forty-seven) is found in western North America from California to Mexico. The few species that reach sub-arctic regions, e.g. Scots Pine (*P. sylvestris*) and Jack Pine (*P. banksiana*), are but a small minority and their strictly needle-like leaves are quite unmodified and closely resemble those of species in the entirely different climates of Mexico and Florida. In temperate regions pines occur on impoverished soils (Jack Pine) — and even on bogs (Scots Pine, Europe; Red Pine, northern U.S.A.) or as pioneers of landslip debris and in abandoned clearings (Pitch Pine, eastern U.S.A.) or as climax forest where fire or wind brings recurrent destruction (Lodgepole Pine, western U.S.A.). Pines

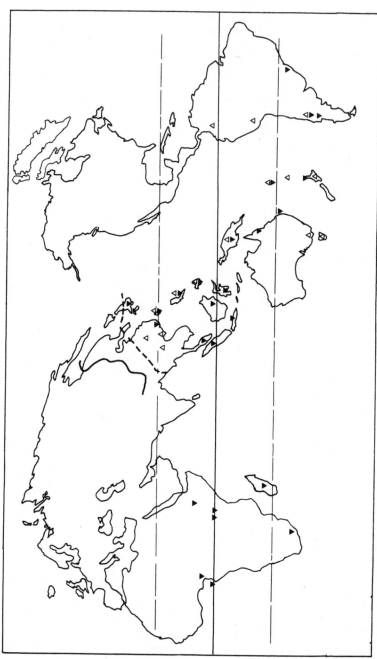

6.2 Distribution of broad-leaved and needle-leaved conifers in the tropics and sub-tropics. Broad-leaved types (▲) include species of *Agathis*, *Araucaria* and *Podocarpus*. Needle-leaved types (△) include *Acmopyle*, *Cunninghamia*, *Keteleeria* and species of *Araucaria* and *Podocarpus* as well as members of the Yew Family. Solid line: northern boundary of *Taxus*. Broken line: northern boundary of *Cephalaxus*. Data from Krussman 1955, van Steenis and van Balgooy 1966 occur in swamps (Pond Pine, *P. serotina*, Atlantic coast plain) and on porous soils, as in the "barrens" of the Carolina piedmont, while in semi-arid regions of warmer climates few angiosperm trees can compete with them. Apparently, then, all pines are equipped with a leaf form which should confer drought resistence though the environments occupied by some species do not demand this capability and possibly some of them do not possess it. In any case many pines owe their success to effective seed dispersal which enables them to colonize rapidly and from a distance the temporary habitats presented by disturbed ground and immature soils. "In regions other than deserts, tundra and the tropical, humid lowlands, pines may be found in nearly any conceivable habitat" (Mirov).

The occurrence of needle-leaved forms in sub-arctic and semi-arid regions can be understood on the grounds that this type of foliage is xeromorphic, that is, it possesses features that one would expect to be effective in reducing water-loss from the plant. If its effectiveness can be demonstrated in practice, then the features of the needle leaf constitute an adaptation enabling plants to live in conditions of at least seasonal drought. The "drought" of sub-arctic regions is described as "physiological" (within the plant) and it occurs when the saturation deficit of the air causes water-loss by transpiration while frozen soil still inhibits water-intake by the root system. However, due regard must be given to examples of plants with virtually the same xeromorphic features of the leaf which never experience such extremes. The flattened needle leaf form of boreal firs (*Abies balsamea, Abies sibirica*) is shared by other conifers such as Hemlocks (*Tsuga canadensis, Tsuga heterophylla*) in North America and Yew (*Taxus baccata*) in Europe, which are distinctly temperate species incapable of growing further north. They endure neither arid summers nor severely cold winters and require year-round precipitation. Likewise, Douglas Fir (*Pseudotsuga menziesii*) with foliage of the same type is the dominant tree of wet temperate forest in oceanic districts of western Canada and U.S.A. Its northern limit fails to reach the boreal zone of southern Alaska. Despite the fact that these trees possess rigid needle leaves they do not tolerate the conditions of permafrost or aridity: indeed, their tolerance in these respects is exceeded by many trees with broad-leaved deciduous foliage. Most of the conifers from central China southward to Tasmania and Chile, whether needle-leaved or otherwise, are not hardy enough to withstand cool-temperate climates and cannot be grown in botanical gardens at Berlin or St. Louis, for instance, without special protection from frost at least. Yet another conifer with foliage of the same general form, though in this case less rigid, is the Swamp Cypress (*Taxodium distichum*) of the south-eastern United States which grows with "its feet in water", i.e. alongside creeks and on inundated ground, in a moist climate in which frosts are short-lived and not severe. Very circuitous arguments would be needed, therefore, to support the claim that the needle leaf form represents an adaptation.

A further point of interest is that in some tropical and sub-tropical genera of conifers the leaves of young plants are needles even when the mature tree has foliage of shorter, imbricate, scale-like form approaching the "cupressoid" type. Juvenile forms are considered by many botanists to represent the ancestral type. In some species both kinds of foliage are normally present on the tree at maturity, the scale-like leaves being characteristic of .the cone-bearing twigs (as in the Redwood, *Sequoia sempervirens*). There seem to be no grounds for arguing that these twigs are adapted to conditions different from those in the other parts of the tree! Surely the needle leaf must be seen as an ancient inheritance of conifers which has little to do with adaptation to environmental severity.

The gymnosperms, in which the conifers are included, are a geologically ancient group, much older than the flowering plants (angiosperms). The internal structure of the wood and the external form of the leaf in living conifers immediately identifies their relationship with fossil predecessors. They have changed comparatively little in morphology over vast periods of time and modifications that appear to convey greater fitness for particular environments are rarely encountered. Surely these cone-bearing trees are the survivors of plants that populated the earth in earlier periods of its geological history. Historically they were overtaken by the rise of the flowering plants whose differing wood anatomy, foliage form (broad and relatively pliable) and mode of reproduction all conveyed advantages in efficiency which led to their displacing the gymnosperms from places where conditions permitted more prolific growth. Thus, through competition as they evolved, flowering plants came to occupy the most favoured situations but their success was not overwhelming in all environments. The gymnosperms maintained themselves in the face of competition on the middle and upper slopes of tropical mountains in forests of mixed composition. On the upper parts of mountains further north and south and in the lowlands of the sub-arctic their more robust form proved to be less susceptible to damage by wind and frost than the more delicate structures of angiosperm trees. Thus, in environments that are less hospitable in various ways gymnosperms remain in occupation un-challenged. It now becomes more comprehensible that gymnosperms can be found in all sorts of harsh conditions in most parts of the world, often not in great variety in such areas and usually not accompanied by many kinds of flowering plants (angiosperms). However, they are by no means confined to rigorous environments and this view of their survival in so many diverse situations is very different from the assumption that their present forms, and particularly their leaf forms, evolved as an adaptive response to the conditions they now experience.

A closer integration of the evolutionary and the ecological view-points in biogeography can enlighten many of the problems that are often discussed in isolation by those interested in the ancestry or genetic lineage of living organisms and by those interested chiefly in the environments they occupy. Certainly in plant geography a greater awareness of the ancestry, the geological history and the natural (phylogenetic) relationships of plants even in the context of the composition of vegetation would lead to a much wider view of the reasons for present existence of particular forms in particular locations.

Spiny plants

The presence of spiny plants in the vegetation of mediterranean and semi-arid climates is another feature of growth-form which in older texts is cited as adaptive. The reason behind this thinking is that the spiny plants in question have reduced the area of the foliar surface by various modifications. Most commonly the leaves are replaced by rigid spines, perhaps derived from the leaf stalk, and in other cases short lateral shoots take the form of spines, as in European Gorse (*Ulex europaeus*). This plant has been mentioned as a sub-atlantic element in terms of its distribution and its natural area of occurrence had its southern limit in northern Portugal. It has been introduced into the Mediterranean region and Rikli classed it as a Mediterranean xerophyte. Clapham (1943) writes, "Its claim to be called a Mediterranean xerophyte must be founded on a demonstration that it can survive when truly exposed to the dry hot summer of the typical Mediter-ranean littoral." Yet, "in the Mediterranean basin it appears to be chiefly a constituent of the undergrowth of woods in moist situations, as, for instance, under *Pinus pinea* near Pisa and Ravenna." Both its natural distribution and its behaviour following introduction into the Mediterranean suggest that it cannot legitimately be classed as a xerophyte despite its suitably modified morphology. Other species of Gorse (*Ulex gallii, U. minor*) have eu-atlantic distributions which tie them to markedly oceanic regions with evenly distributed rainfall throughout the year and a general absence of even temporary drought, despite their possession of the same life-form and xeromorphic features.

Genista corsica is a spiny plant whose distribution is Mediterranean and which may justifiably be called a xerophyte, but among closely related species *Genista anglica* has a eu-atlantic distribution although possessing the same spiny characteristics of form. "*Genista anglica* is one of the most atlantic species of our flora together with *Myrica gale, Ulex europaeus, Hypericum elodes, Erica cinerea* and *Erica tetralix*" (Hegi in *Illustrierte Flora von Mitteleuropas*). *Genista anglica* would be most inappropriately regarded as a xerophyte for its preferred habitat is in bog communities where the water-table is never more than a few centimetres below soil surface and this in a humid climate! Evidently certain morphological characteristics which were previously thought to have adaptive significance are better regarded as family or generic traits exhibited by many species simply because they neither enhance nor detract from the functional effectiveness of the plants that carry them. A human analogy would be that a majority of members of a Scottish clan have red hair but this characteristic is not abandoned when certain member families emigrate to settle in other climates nor does it change in relation to the job or profession of the individual.

The ericoid leaf

The fallibility of a functional link between many details of plant form and environmental conditions can be illustrated again by the ericoid leaf, a distinctive type in angiosperm shrubs, dwarf shrubs and perennial herbs. It is comparable to the needle-leaf of spruces and other coniferous trees. The narrow leaf has its outer edges rolled inwards to produce two grooves along the underside. The lower surface of the leaf is usually clothed with hairs and the stomata (the pores through which gaseous exchange takes place) are confined or concentrated in the grooves. These particulars of leaf structure are all considered to be xeromorphic, i.e. they should be expected to reduce the loss of water by transpiration through stomata which are thus protected. This expectation seems to be fulfilled from the circumstantial evidence that the heath family (Ericaceae) is represented by a profusion of species (c. 500) in the Cape region of South Africa where an arid summer is characteristic of the climate. In Europe the species of this family, though few in number, tend to dominate certain types of vegetation — heath and moor — and are therefore conspicuous in the landscape. Remarkably this is true only of western Europe where oceanic influence is a strong element of the climate. The total distributions of these European species are eu-atlantic to sub-atlantic e.g. *Erica ciliaris*, *Erica cinerea*, *Erica tetralix*, and so they do not experience a climate in which resistance to drought is a necessity. Indeed, *Erica tetralix* is confined to bog habitats where there is constantly moist organic soil, and others such as *Calluna vulgaris* are almost equally characteristic in the same situations. Clearly, in some cases their xeromorphy does not indicate their physiological tolerance and hence their ecological character. The point emerges that in plants appearances may be deceptive. This is not to deny that in many cases attributes of form may be effective in aiding the plant's survival under severe conditions but possession of those attributes is not always confined to species in which it has functional significance. Possession of xeromorphic features is not a reliable indication of the climate to which a plant is adapted. In any case the operative features are often more minute in scale than the characteristics which are used in describing life-form or the physiognomy of vegetation. For instance, the degree of hairiness of the leaf-surface and the fact that a grass leaf may be rolled (when seen in cross-section) are much more likely to have significance than any gross features of the plant form. It would appear that characteristics evolved in an earlier stage of the group's history, which perhaps had survival value then, are frequently retained (not lost) in descendant species now inhabiting quite different conditions, providing those characteristics do not constitute any disadvantage to the plant's functions.

Sclerophyll plants

The example of the sclerophyll leaf reveals similar anomalies and also shows that it may be difficult to decide where to draw the line. Since we have to observe clear definitions in any code of life-form description we shall find included in our categories species whose geographical distribution seems to conflict with the supposed adaptive significance. A sclerophyll is a type of leaf usually of small size, characterized by its tough or hard texture and its persistence on the shoot for more than a single season (commonly for 2-3 years). Plants which possess sclerophyllous foliage are therefore evergreen. In applying this definition there comes a point when it is questionable whether certain species properly qualify for inclusion. Thus undisputed examples in Europe are Holm Oak (*Quercus ilex*) — south European, Holly (*Ilex aquifolium*) — sub-atlantic, Olive (*Olea sativa*) — Mediterranean, Myrtle (*Myrtus communis*) — Mediterranean, and in North America Creosote Bush (*Larrea divaricata*), *Ceanothus* species and Manzanita (*Arctostaphylos*). Clapham (1943) questions whether Strawberry Tree (*Arbutus unedo*) can be usefully placed in the same class as sclerophylls. Its leaves are small, they are persistent and they do have the typical glossy upper surface due to a thick cutinised coating. But the leaf texture is not markedly tough or leathery: most people would probably describe it as papery. Is this to be classed as a sclerophyll? Moreover "Do the features by which they are placed as sclerophylls have any real bearing on their capacity to thrive as ecologically important members of Mediterranean vegetation?" (Clapham).

Sclerophylls are also prominent in the mixed evergreen forests of the Coast Ranges of Pacific North America from Washington to north California where many species of shrubs such as Barberry (*Mahonia* spp.) and Manzanita (*Arctostaphylos* spp.) form an understorey beneath evergreen trees, chiefly oaks and Douglas Fir. Foliage of the same type is equally important in the Sand Pine scrub of Florida and south-east Georgia. Here a dense vegetation of hard-leaved shrubs 1-3 m high includes several species of scrub oaks, two dwarf palms and the ericaceous shrubs *Vaccinium arboreum* and *Lyonia ferruginea*. In the mountains of Virginia a sclerophyll scrub vegetation is formed by *Kalmia latifolia* (also a member of the family Ericaceae). There is no obvious aspect of climate common to these three situations and none of them are of the "mediterranean" type, with which sclerophyll vegetation is popularly associated.

Possession of a particular life-form does not restrict a species to that environment in which the peculiarities of its morphology are supposedly adaptive. Some of the species discussed as examples are not excluded from other climates than those of the areas mentioned. Holly is a common tree throughout the British Isles and Strawberry Tree occurs naturally in the highly oceanic woodlands of County Kerry, south-west Ireland. Likewise, its relative in North America, Madrone (*Arbutus menzesii*) extends northward to the humid temperate forests of Vancouver Island. Since no single life-form exists in obligate relationship with a particular climate we find representatives of several or many life-forms in any piece of vegetation. The degree of association between life-forms and climate can only be expressed therefore in terms of the proportionate representation of life-forms among all the species present. Alternatively the order of importance of the life-forms can be assessed by their abundance among individuals of all kinds, regardless of species. In this procedure inferences are drawn from the behaviour of the majority but the success of species belonging to life-forms that are in a minority is in no way explained.

Evergreen and deciduous plants

With reference to sclerophylly, in the south-east United States, Monk (1965) has shown that percentage of evergreens (here including many sclerophylls) increases on drier, more sterile sites, while predominantly deciduous vegetation occupies the more mesic, fertile sites. Following an analysis of some environmental factors with which a statistical correlation can be shown, Monk (1966) concludes that evergreens may conserve their resources of mineral nutrients better than deciduous species and this could be responsible for their prevalence on improvished sites. This is not a region in which there is deficiency of rainfall so that the presumed resistance of the sclerophyll to desiccation does not seem relevant in this case. The advantages they possess here may operate through the functional attribute of being evergreen rather than through any anatomical feature of their leaf form, e.g. sclerophylly. It is established that nutrients are leached by rainfall from the leaves on the plant and thus returned gradually in small amounts to the soil. In an evergreen this can be effective throughout the year. Moreover the fact that dead leaves are shed from the plant in small numbers continually round the year likewise returns nutrients to the soil regularly instead of just once a year in massive quantities as in deciduous vegetation. In the latter case the soils would tend to lose much of this returned nutrient material by leaching, especially in regions of moderate to heavy rainfall and on porous soils. Monk does not exclude the possibility that in semi-arid and arid climates sclerophyll plants may owe their existence to water-conserving ability.

Clearly there is a strong possibility that the same life-form may be effective in ensuring survival by quite separate aspects of function in the various climatic regions in which it occurs. The value of referring to this as "adaptation" is doubtful: it is liable only to give a false impression and it would differ from the usage of this word as applied in the animal world.

Functional periodicity

Finally, we must consider the periodicity of functions in evergreen and deciduous plants. Too often the deciduous habit is associated with temperate seasonal climates in which a period of winter cold brings to an end the season of growth. It is also characteristic of seasonal climates in lower latitudes in which a period of aridity is unfavourable to plant growth. Here seasonal fluctuation in the water-balance parallels the seasonal temperature change of the northern latitudes in inhibiting plant growth. In contrast it is commonly believed that in ever-wet non-seasonal climates of the tropics growth is continuous. Both these ideas contribute to an impression of plant life as a mute and plastic "response" to the exigencies of climate, and therefore as a living index of climatic type.

Allowing that the vegetative period of deciduous plants varies in duration from place to place, we must recognise that the strict definition of a deciduous plant is that it is a plant which sheds all its foliage within a short period and stands bare of leaves at some time in the year. The evergreen plant, of course, also sheds its foliage, since the individual leaves do not remain in position for the entire life of the plant. The difference is that the leaves of an evergreen are shed more or less continually as they reach the end of their functional life but at no time is the plant entirely without foliage.

Some observations by Njoku (1963) in southern Nigeria are informative in this connection. Ten species of trees that occur in tropical lowland rain forest were carefully observed over five years and their periodicity in growth was recorded. Some of the species chosen were deciduous and others evergreen but all of them without exception show a marked annual periodicity in function. The several functions recorded were leaf-fall, flowering and state of the vegetative buds, either active or dormant. The buds give rise to new leaves and it is important to know that in these species new leaves are not continuously unfolding but are produced only at specific times in the year when the buds are active. The number of additional leaves produced on each shoot per year is limited within a narrow range for each species and may be as few as two. The vegetative buds are responsible for extension in length of the shoots so that in this sense also growth is distinctly periodic, occurring only when the buds are active. The description "buds dormant" means that additional growth in shoots and leaves was not taking place: it does not imply that the tree was bare of foliage. The renewal of bud activity following a period of dormancy is termed bud-break.

From the periodic behaviour of these species (Fig. 6.3) we come to realize that the distinction of deciduous and evergreen in tropical areas is something of a technicality, although of course the deciduous habit is conspicuous in its effect. Nevertheless a tree which has compressed its leaf-fall into one month of the year instead of carrying on this process imperceptibly through all twelve months becomes technically deciduous if leaf-fall is completed before bud-break renews the foliage, even if it remains bare for only 2-4 weeks. Because there is some variation in phenology from year to year, in the extreme case, e.g. *Bosqueia angolensis,* in which leaf-fall and bud-break often occur within the same month, there are some years in which the tree is evergreen: the new crop of leaves is already expanding before the previous year's crop is shed. In contrast, by extending the period between leaf-fall and bud-break, the condition of deciduous trees in temperate latitudes is attained. *Bombax buonopozense* exemplifies this situation in Nigeria, the tree being devoid of foliage for 3-4 months.

In addition to recognising that evergreen trees have periodicity and that in some at least this includes growth and leafing, not merely flowering (which is periodic in all), we must consider the incidence of these events in relation to environmental conditions. Ibadan, where Njoku's study was carried out, is on the northern edge of the rain forest zone and therefore is in a critical position to register the effects of seasonality. In most of West Africa the forest exists in a climate with well-marked dry and rainy seasons. The number of "dry" months (recording less than 10 cm rainfall per month) increases northward away from the coast and at Ibadan the dry season extends to six consecutive months thus defined. Despite this marked seasonality the behaviour reported for these ten species is not very different from that of similar species growing in the wetter forests further south where only 2-3 "dry" months are experienced, none of these having less than 2.5 cm of rainfall. In other words, the timing of these periodic events in

	NOV.	DEC.	JAN.	FEB.	MAR.	APR.	MAY	JUN.	JUL.	AUG.	SEP.	OCT.

6.3   Periodicity of functions in ten Nigerian forest trees. Data from Njoku, 1963.

tree growth is not altered by relative lengths of wet and dry seasons. Moreover, at Ibadan itself the periodicity in plant function differs between one species and another (under the same climate) and in none of them is there any correlation between vegetative activity and rainfall. The rainy season extends from April to September and Njoku points out that in most of the examples cited bud-break occurs in•February and March. Since leaf production is limited to about one month in the year there is no further addition of foliage (only extension growth) during much of the rainy season when water supply is plentiful. Moreover, leaf-fall in the deciduous species occurs in December-January, about half-way through the dry season, and as these are leafless for only a small part of the dry season, he concludes that these foliage changes "have no adaptive value in enabling trees to avoid excessive transpiration" in the period of relative dryness.

This instructive observation shows that even in a strongly seasonal forest climate the deciduous habit is not ubiquitous: some evergreen species persist especially among trees of the B storey in stratified forest. Furthermore the deciduous species do not become clothed with leaves or defoliated synchronously with the change of seasons. So short is the leafless period that adaptation cannot be invoked.

Corner (1955) believes that all patterns and variations of plant behaviour co-exist in the tropical rain forest (as we have seen) because it is here that they have evolved at random, unrestrained by selective pressures. In so favourable an environment for plants, never lacking moisture or warmth, all variants may find a place and thus survive. In his view those that succeed in tolerating less favourable conditions, e.g. diurnal and seasonal fluctuations, extremes of temperature and water-supply, an annual cycle of day-length and light supply, etc., do so by virtue of the inherited properties that they already possessed and not by special adaptive evolution in response to those environments. Throughout this account we have seen good reason for finding this interpretation acceptable. The facts that many life-forms are capable of co-existing in most situations and that each of them finds itself suited in some way or another to existence in a variety of environments (in the guise of different species) suggest that they are not purpose-made adaptations. In using the life-forms of plants for describing and classifying vegetation it will be advisable to discount all preconceived ideas on their adaptive value. Providing a species is not positively unsuited to survive in a certain environment by the life-form it possesses, it may grow there limited only by functional conditions beyond its tolerance. In many cases external form belies the real capacity of a species to thrive under the most varied circumstances.

SOURCES OF REFERENCE

Bews, J.W. (1925) *Plant Forms and their Evolution in South Africa*. London.
Clapham, A.R. (1943) Book review. *New Phytologist* vol. 42, pp. 59-62.
Corner, E.J.H. (1955) *The Life of Plants*. Weidenfeld and Nicholson, London.
Dallimore, W. and Jackson, A.B. *A Handbook of Coniferae*. 4th edn. revised by S.G. Harrison (1966). Arnold, London.
Krussman. G. (1955) *Die Nädelgehölze* Paul Parey, Berlin and Hamburg.
Mirov, N.T. (1967) *The Genus Pinus*. Ronald Press, New York..
Monk, C.D. (1965) Southern mixed hardwood forest of north-central Florida. *Ecological Monographs* 35, pp. 335-54.
Monk, C.D. (1966) An ecological significance of evergreenness. *Ecology* 47, pp. 504-5.
Njoku, E. (1963) Seasonal periodicity in the growth and development of some forest trees in Nigeria. *J. Ecol.* 51, pp. 617-24.
Richards, P.W. (1957) *The Tropical Rain Forest*. Cambridge, Cambridge University Press.
Rikli, M. (1942-3) *Das Pflanzenkleid der Mittelmeerländer*. Hans Huber, Berne.
Steenis, C.G.G.J. van and Balgooy, M.M.J. van (1966) Pacific Plant Areas. Vol. 2 *Blumea* Supplement 5, Leyden, Netherlands.

CHAPTER 7

Vegetation: classification and correlation

Introduction

A number of characteristics can be used as a means of classifying vegetation, among which structure, function and composition are the most important. Composition may be treated ecologically in terms of life-forms or floristically in terms of species. The objects of classification are to create a systematic framework for ease of reference and to achieve that measure of generalization that will enable us to compare geographically distant vegetation types and perhaps associate those types with particular geomorphic or climatic regions. Classification is also necessary before mapping of vegetation can be attempted.

Most reference atlases include a map of world vegetation though at the scale it is practicable to publish these maps the information they convey is about as useful as world geology shown at the same scale. It is at best a very imperfect guide to what kind of vegetation-landscape the traveller might expect to see in different countries on his itinerary. The limitation of scale is one of the reasons for the inadequacy of such maps because obviously the actual vegetation is a mosaic of plant cover reflecting the intimate variations in ground conditions — slope, drainage, soil depth, exposure — and only the most extensive type can be represented. This is complicated by the fact that most vegetation maps display the original vegetation type of each region and not the man-modified plant cover which is often more extensive today and largely determines the present landscape character. However, this is not a criticism for there is great theoretical interest in knowing the ultimate natural vegetation of an area since to a large extent it reveals the productive capability of the land under the conditions of climate prevailing. This refers to production in natural ecosystems which are by definition balanced in the sense that plant and animal produce is regulated at a level that the mineral nutrient and water resources can support indefinitely. Potential for human exploitative production may of course be greater in artificial systems of land use management.

A more serious cause of inadequacy in many world vegetation maps is the classification adopted. This inadequacy arises from the use of non-vegetational criteria, especially climatic features, in naming the units recognised. Thus "rain-forest" may be a useful term in describing the vegetation of Brazil when its local connotation has been explained but at world scale it conveys only that the vegetation is predominantly a closed canopy of trees in a wet climate. The further distinction of tropical rain forest and temperate rain-forest does nothing to improve our knowledge of the vegetation because it relies upon latitudinal or climatic separation and any vegetational difference is merely inferred. "Desert" is a term of uncertain meaning, being a word used in everyday language as well as in different aspects of geographical study. Unless it is first defined, its use in a vegetation map may be conditioned by ideas on what is climatically a desert or what is geomorphologically desert terrain. Interpreted literally it means devoid of life and this would exclude many areas to which the name is habitually applied. The important point is that at face value the word does not provide a description of vegetation in such a way that it can be visualised in its general character.

Structural classification

The first requirement for a satisfactory classification of vegetation is that it should be entirely founded on plant characteristics and should not involve any dependence on environmental or locational factors. The most suitable criteria are therefore the structure and the functional character of the vegetation. Of these, according to Dansereau (1957), "structure . . . is one of the outstanding features of vegetation and ranks even before composition in a description of landscape." He has proposed ten types of plant formation which are comprehensive and of universal application. These are named: (1) forest, (2) woodland (*parc*), (3) savanna, (4) scrub (*fourré*), (5) prairie, (6) meadow (*pelouse*), (7) steppe, (8) desert, (9) tundra, and (10) *croûte*, freely translated as mat, a very low-growing vegetation cover. In the proposed usage these terms are strictly to describe the spatial arrangement of plants comprising the vegetation, particularly in regard to their height and distance apart. Thus, for example, a strict definition is given for the word "desert". While it is apparent that these formations are listed approximately in order of diminishing stature it should also be noted that there is a continuous gradient in the density of plants per unit area within each height layer, e.g. from forest to savanna, and from prairie to steppe. In nature there are no distinct breaks between one type and the next. This explains the need for the arbitrary but definite limits which Dansereau has prescribed (Table 3). It is important to appreciate

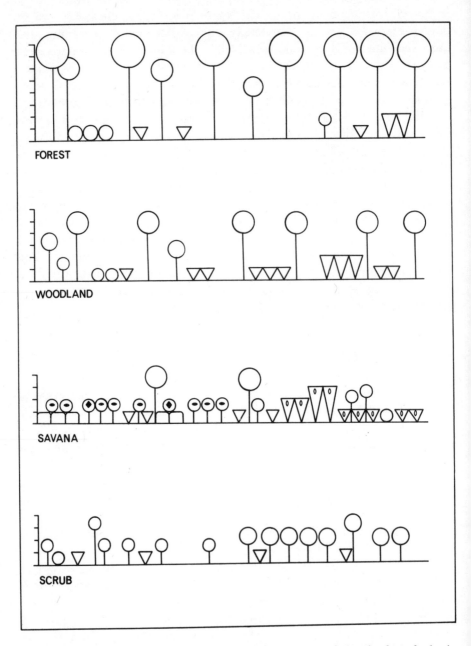

7.1 The ten primary vegetation formations of Dansereau, determined on the basis of structural characteristics, viz. stratification and spacing. (Adapted from Dansereau & Arros, 1959.)

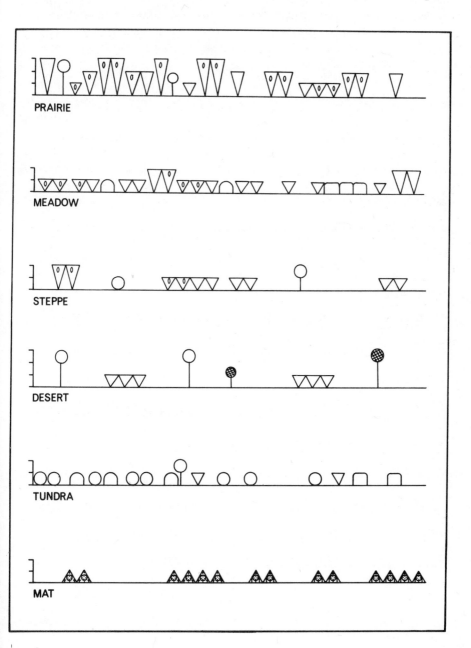

PRAIRIE

MEADOW

STEPPE

DESERT

TUNDRA

MAT

that these ten types of formation are solely structural: they do not invoke or assume any other influence, floristic, climatic or edaphic. It is intended that "savanna" may be used for the appropriate formation of plants in any area, in any climate. In this scheme it does not imply that the vegetation necessarily includes plants of the grass family, or that it experiences a dry season or is rooted in a deep soil. Some of these things may be true, of course, in particular areas, but they are not implicit in the terms and the association should not be assumed. Allowing for variations that occur in nature, each formation embraces a range of structural profiles fundamentally the same in height and stratification but varying in other respects (e.g. density) within certain limits, as illustrated by the life-form symbols and their spacing in Fig. 7.1.

Table 3   The formation classes of Dansereau

	Stratification	Cover Value	Cover Description
Forest	Woody pls. >8m	Cover >60%	(c)
Woodland	Woody pls. >8m	Cover 25-60%	(i-c)
Savanna	Woody pls. 2-10m herbc. pls. 0-2m	Cover 10-25% Cover 25-100%	(i-p) (i-c)
Scrub	Woody pls. 0.1-8m	Cover 25-100%	(i-c)
Steppe	Woody pls. 0.1-2m herbc. pls. 0-2m	Cover 0-25% Cover 10-50%	(b-i) (i-p)
Desert	Woody pls. 0-10m herbc. pls. 0-0.5m	Cover 0-10% Cover 0-10%	(b) (b)
Tundra	Woody pls. 0-0.1m herbc. pls. 0-0.1m	Cover 10-60% Cover 0-20%	(b-c) (b-p)
Prairie	herbc. pls. 0-2m	Cover 50-100%	(c)
Meadow	herbc. pls. 0-0.5m	Cover 50-100%	(c)
Mat	herbc. pls. 0-0.1m	Cover 50-100%	(p-c)

Key to abbreviations:

    c   continuous   i   interrupted
    p   scattered *(parsemé)*   b   sparse (barren)

Herbaceous is used here in an extended sense to include all non-woody plants: herbs, grasses and bryoids of all kinds.

A classification proposed by F.R. Fosberg (1967) and adopted for general purposes by the International Biological Programme also regards structure as fundamental but differs slightly in emphasis. The horizontal spacing of plants is regarded as a primary feature and three "primary structural groups" are distinguished: (i) closed vegetation, in which the plants are predominantly touching or overlapping (either within any one layer or in all the layers taken together); (ii) open vegetation, in which neighbouring plants are not touching but are separated by not more than their diameters; (iii) sparse vegetation, in which the distance between plants is greater. In order to qualify as "open" (ii) all the layers in a multi-storeyed vegetation must meet the criterion given above. For example, in the savanna type of Dansereau (Fig. 7.1c) the tree layer is "sparse" but the understorey is continuous so that the vegetation is classified as "closed". Fosberg makes category (iii), in which all layers of the vegetation are "sparse", synonymous with desert. His point of view is that vegetationally desert is any situation in which the plants are scattered or separated by intervals greater than their crown diameter (spread), regardless of their height or stature. In Dansereau's scheme this would also apply to steppe which he distinguished from desert by the greater height and cover of herbaceous plants. Fosberg too recognises steppe as "open" vegetation but sets a limit where plants are separated by more then their spread. The demarcations selected by the two authors are based on different criteria, Fosberg's with rather strict reference to plant spacing, Dansereau's with primary emphasis on stratitification and cover combined. Fosberg's three primary structural groups based on spacing are sub-divided into *formation classes* (Table 4) according to the height of the most continuous layer.

Table 4   The formation classes of Fosberg

1 Closed Vegetation	2 Open Vegetation ("steppe")	3 Sparse Vegetation ("desert")
A. Forest	A. Open Forest (Steppe Forest)	A. Desert Forest
B. Scrub	B. Steppe Scrub	B. Desert Scrub
C. Dwarf Scrub	C. Dwarf Steppe Scrub	C. Desert Herb
I. Tall Savanna	D. Steppe Savanna	Vegetation
J. Low Savanna		
	(an extract from the full classification).	

These categories are further sub-divided into *formation groups* by consideration of function, either evergreen or deciduous (always referring to the tallest stratum present), and the ultimate division into *formations* follows by reference to predominant life-form using such characters as leaf-size (microphyll, etc.) shape (needle or broad) and texture (sclerophyll, succulent, etc.).

Despite its logic, the I.B.P. scheme does not easily digest the complexities caused by stratification. We are forced to accept *classes* named, for example, "open dwarf scrub with closed ground cover" and "closed scrub with scattered trees", both distinct from savanna and of equivalent standing to the classes named in Table 4. Furthermore, the use of "desert" as synonymous with distant spacing and of "steppe" with open spacing of plants results in some curious combinations of terms such as "Desert Forest" and "Shrub Steppe Savanna" which simply defeat the object of conveying a readily understood impression. It is not a lack of precision but the use of vernacular terms to which highly special meanings have been given that is liable to cause confusion. In practice, by the use of a key provided, any piece of vegetation could be successfully classified on this scheme. However, it seems regretable that technical terms describing leaf size, such as microphyll, mesophyll and megaphyll, were introduced into this classification while not adhering to the dimensions prescribed in Raunkiaer's definitions and used by other authors. It is also confusing to find that microphyllous is used for three different sizes of leaf when applied to trees, shrubs and dwarf shrubs respectively. The I.B.P. classification draws attention to the relevance of spacing in vegetation but in elevating this to the status of a primary diagnostic feature and adopting over-precise meanings for common terms it has some features that might deter many potential users.

Functional classification

An alternative approach to the classification of vegetation has been developed by A.W. Küchler who proves the scheme practicable in his *Potential Natural Vegetation of the Conterminous United States* with accompanying map at the scale of 1 : 3 million. Function and foliage form are first considered in devising the major groups of vegetation types, thus stressing the fundamental importance of these characters and their visual impact on the physiognomy of vegetation. Ten categories are formulated on this basis, namely:

B.	broadleaf evergreen	O.	without leaves
D.	broadleaf deciduous	G.	graminoid (grasslike)
E.	evergreen needle-leaf	H.	herbaceous
N.	deciduous needle-leaf	L.	lichens and mosses
S.	semideciduous (B + D)		
M.	mixed (D + E)		

The classification describes all possible vegetation forms because these primary terms can be used in any combination to denote the principal layers and their predominant life-forms, which are usually different. The terms can be represented by their code letters to produce an abbreviated description or formula. The convention adopted is that the layer having most extensive coverage within the vegetation is stated first. Density or spacing of plants is indicated by use of Dansereau's grades: continuous (c), interrupted (i), scattered (p), and sparse (b). By way of example the Brazilian llanos is denoted as GSp, that is dominantly graminoid vegetation (G) with scattered broadleaved trees, of both evergreen and deciduous species. It is a simple matter to develop a classification of all types of vegetation by grouping together those which possess the same life-form as their most extensive layer.

For instance, grass-dominant vegetation would link together prairies, steppes and savanna and in cartography these could be designated by the same base colour with variations in tone and line shading.

Subdivision of the broad categories is accomplished by describing the height or stature of the layer referred to. For example, broadleaved evergreen woody vegetation (B) can be sub-divided into tall forest (t), medium forest (m), low forest (l), scrub (s) and dwarf scrub (z). In practice the "medium" category can be neglected since it can be assumed unless stated otherwise, using (t) for forests over 25 m in height and (l) for those lower than 10 m. The same letters apply to graminoid and herbaceous vegetation in which case (t) denotes plants more than 2 m high and (l) less than 0.5 m. Thus Gt represents tall prairie and Gl is short prairie; GtDp being tall grass savanna (i.e. with scattered deciduous trees). The vegetation of the Kara Kum desert is mapped as Dsp, indicating a scattered scrub of deciduous broadleaved woody plants. Already in four digits the classification informs us of principal life-forms, layering (stratification) and spacing (density, cover).

This does not exhaust the possibilities of Küchler's classification for it provides a framework for classifying individual plant associations named by their principal species. In effect subdivision below the level of four digits is based upon floristic composition. For the sake of completeness, the associations included in the *Broadleaf Forest* formation of the United States are listed to illustrate the floristic subdivisions adopted by Küchler (1964).

Northern floodplain forest (*Populus — Salix — Ulmus*)
Maple — Basswood forest (*Acer — Tilia*)
Oak — Hickory forest (*Quercus — Carya*)
Elm — Ash forest (*Ulmus — Fraxinus*)
Alder — Ash (*Alnus — Fraxinus*)
Beech — Maple forest (*Fagus — Acer*)
Mixed mesophytic forest (*Acer — Aesculus — Fagus — Liriodendron — Quercus — Tilia*)
Appalachian oak forest (*Quercus*)
Oregon oak woods (*Quercus*)
Mesquite bosques (*Prosopis*)
Mangrove (*Avicennia — Rhizophora*)

The merit of this procedure is that it acknowledges that the plant associations of each geographical region are unique and does not attempt to make the finer distinctions in general terms. Now that the vegetation of the world can be classified objectively without resort to conditions of environment or location, we can compare the vegetation types of regions that are similar in climate and examine the geographical occurrence of vegetation types that are comparable in life-form, structure and function.

Humid temperate climates and forest types

The temperate climates of the world as defined by Miller, with mean temperature of the coldest month not exceeding 68°F (20°C) are divided into warm temperate (winter months >43°F (6.1°C)) and cool temperate, in which at least one month has a mean temperature below 43°F. Excluding regions of aridity (which are separately classified) these cool temperate climates have adequate rainfall for forest formations. The regions in which they occur in the northern hemisphere (Fig. 7.2) are (i) pacific North America from latitude 40° to 60°N (Oregon to Juneau, Alaska); (ii) atlantic Europe from 42° to 62°N (Bilbao to Bergen); (iii) eastern U.S.A. and southern Canada (St. Louis to Quebec); (iv) central Europe (Budapest to Leningrad); (v) central Japan and adjacent east Asia. Since the climatic boundaries to these regions are arbitrary, no one would expect forest boundaries to be coincident with them, but what is perhaps surprising is the variation in dominant forest type between the separate areas. The first two regions present the greatest contrast since on Küchler's world vegetation map (in Goode's Atlas and Polunin 1960) pacific North America is dominated by evergreen needle-leaf forest of Douglas Fir, Sitka Spruce and Western Red Cedar while atlantic Europe is (or was) dominated by deciduous broad-leaved forest of Oak, Beech and Hornbeam. If we are wary of facile "adaptive" correlations, it is clear that the biogeographic histories of these two regions account for their nonconformity in vegetational cover. It means simply that after the most recent catastrophic event in the earth history of these areas, namely the final glaciation of the Quaternary era, different opportunities for migration existed in the two continents. Forest-forming trees of deciduous habit

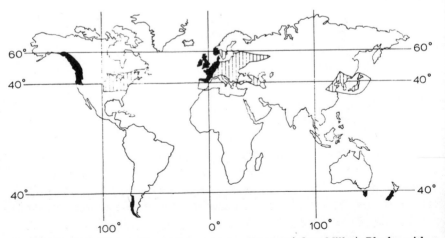

7.2 Distribution of humid cool temperate climates (after Miller) Black: with a short cold season (1-5 months with mean temperature below 43°F (6°C) |||| : with a long cold season (6 months or more with mean temperature below 43° F).

were available to recolonize western Europe from adjacent non-glaciated areas whereas pacific America was not geographically accessible to such trees at that time and their place was taken by the needle-leaved conifers that were available.

Comparison of the eastern U.S.A. with central Europe, which is climatically similar (e.g. Chicago and Warsaw) produces a more orthodox result. Deciduous broadleaf forest prevails with outliers of boreal needle-leaved forest in less favoured situations. The cool temperate region of Japan, the island of Honshu, adds the evergreen broadleaved forest to the range of vegetation encountered in these climates. Before agriculture this was the prevalent type in the southern half of the country, south and west of Tokyo. This emphasizes that the precise character of the vegetation found in a particular region depends upon the species that are tolerant of its climate and available to populate it. It also depends upon the history of events: whether volcanic activity, glaciation or marine submergence have destroyed an older vegetation and if so, what routes have been open and what source areas accessible to replenish the flora and fauna.

The southern hemisphere cool temperate climates all lie just south of latitude 40°S in Tasmania, South Island of New Zealand and southern Chile. Despite their climatic affinity with pacific north-west America and atlantic Europe, their forest vegetation resembles neither of these. It is closer in form to that of Japan, being broad-leaved evergreen, yet again quite different in composition, with its abundance of distinctive conifers. Recalling earlier discussion, many of these qualify as broad-leaved trees (e.g. *Agathis, Podocarpus* Fig. 6.1) while many others have cupressoid foliage (e.g. *Fitzroya, Libocedrus, Widdringtonia*). The index trees of these southern forests are the southern beeches (*Nothofagus*)

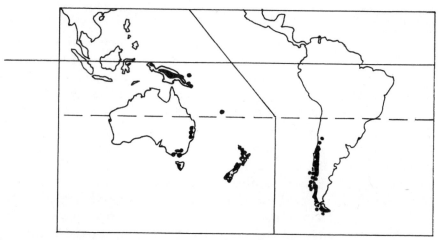

7.3 World distribution of Southern Beech (*Nothofagus*), characteristic of forests in humid cool temperate climates of the southern hemisphere (at high elevations in low latitudes). After van Steenis & van Balgooy, 1966.

(Fig. 7.3) represented by many species, most of them evergreen unlike their northern hemisphere counterparts (*Fagus*). Here again, therefore, we see that dissimilar forms can dominate the vegetation in areas climatically alike but geographically separate.

Raunkiaer analysis

The preceding classifications depend on reducing vegetational description to some common denominator such as structure, or leaf characters of the dominant plants. Another possible basis for class-ification depends on relating vegetation types with similar life-form spectra. The life-form analysis of any piece of vegetation in the terms devised by Raunkiaer is an alternative means of characterizing its essential physiognomy, for his categories uniquely summarize both the general appearance of plants as well as much of their functional behaviour. In the use of this method we are thinking particularly of an inventory of all the species present allocated to the appropriate life-form class. From this a formula known as the life-form spectrum is derived in which the percentage of species belonging to each class is stated. A classification can be drawn up by relating all vegetation types that show quantitative similarities in their life-form composition and this can be made hierarchical by selecting various degrees of similarity. Probably because a complete knowledge of the species present is required for all vegetation types to be included, no general classification based on this principle has as yet been developed, so far as the author is aware. It could prove to be extremely informative because it takes account of the floristic composition of vegetation, i.e. how many species and of what forms, in a way that none of the existing classifications do.

Few individual plant associations have been analysed by this method: more often the flora of a whole region has been treated. Examples of spectra representing large territories, e.g. entire states, within the area of eastern North America that experiences humid mesothermal and microthermal climates (Köppen) are given in Table 5. A high degree of consistency is revealed but since each spectrum represents all possible habitats within the state it would require a very drastic change in the vegetation of any one of them to alter the resultant spectrum significantly. This use of the method is therefore a very blunt instrument but it can be correlated with the climatic region in the broadest terms. Thus, the cool to warm temperate climates with abundant precipitation are characterized by the predominance of hemicryptophytes with phanerophytes and cryptophytes about equal in second place.

Table 5   Some life-form spectra for eastern North America

Flora and author Number of species	Phanero- phytes	Chamae- phytes	Hemi- crypto- phytes	Crypto- phytes	Thero- phytes
Alabama, Ennis, 1928					
2,012 species	17.0	3.1	47.8	17.1	14.4
Mississippi, Ennis, 1928					
1,724 species	17.7	3.1	49.4	16.2	12.8
Great Smoky Mountains					
1,142 species	19.5	1.7	52.1	15.1	11.5
Connecticut, Ennis, 1928					
1,453 species	15.0	1.9	49.4	21.7	11.7
Indiana, McDonald, 1937					
2,109 species	14.3	1.9	49.0	18.0	16.7
Iowa, Ennis, 1928					
1,320 species	14.8	1.0	48.6	20.9	14.2

Altitudinal comparison of life-form spectra

The spectra for individual vegetation formations and even for particular associations within them will be more informative and probably reflect differences in climate with greater subtlety. Within the Great Smoky Mountains National Park two altitudinally distinct forest types (Table 6) have been analysed by Cain (1945). The cove hardwood forest of the valleys can be contrasted with the flora of the highest altitudinal belt, approximately from 4,500 ft to 6,500 ft elevation, which corresponds in general with the subalpine forest dominated by Red Spruce and Fraser Fir. It should be acknowledged that while the cove forest was sampled intensively using circumscribed plots the subalpine data result from extensive recording and are not confined strictly to forest plots of the Spruce-Fir association. This accounts for the misleading impression that the high altitude forests contain more species than the low altitude forests, which is not true. Admitting that differences in data collection could give rise to discrepancies in results, the very strong contrast shown by these two spectra seems to exceed in magnitude the variation that might be caused artificially and it is therefore treated as significant.

Table 6   Altitudinal comparison within Great Smoky Mountains National Park (from Cain, 1945)

	No. of species	Phanero- phytes	Chamae- phytes	Hemi- crypto- phytes	Crypto phytes	Thero- phytes
Cove hardwood forests	113	36%	4.4%	30%	26%	3.4%
Subalpine belt	301	21%	2.3%	56.5%	17%	2.6%

The percentage of phanerophytes is much greater in the cove forest, an association that is very rich in tree species. Of the 64 species of phanerophytes listed from the subalpine belt as a whole only 20 are trees exceeding 8 m in height. Conversely the subalpine belt shows a greater proportion of hemicryptophytes, which gain protection in winter from snow cover. Geophytes, on the other hand, are better represented in the forest at low altitude so that their interpretation as winter-adapted forms seems questionable. More likely their abundance is related to the fact that cove forest includes mostly deciduous trees and therefore presents a spring aspect of high light incidence which does not occur under the evergreen canopy of spruce and fir.

The differences between the two spectra can be safely correlated with differences in the climate conditions they experience since both refer to the same National Park and are not separated by considerable horizontal distance. The climatic change with altitude chiefly affects temperature range, frequency and duration of frosts and length of snow lie.

Latitudinal comparison of life-form spectra

If we compare several American deciduous forest associations by their life-form spectra we see how the unique character of every individual forest is faithfully recorded in the percentage values. The circumstances of geographical location, ecological history and never identical combination of climatic conditions ensure that even the forests of adjacent states have individual differences within the context of their overall similarity. As a means of classification, therefore, the life-form spectrum differs from other methods described in that it never places quite the same label on any two types of vegetation. It is unlikely that the proportional values of all life-forms will be precisely the same even in the corresponding vegetation of two adjacent states. The approximation of these values, however, indicates the degree of resemblance between the types compared.

Interesting possibilities for distinguishing vegetation types are opened up by this method. In Table 7 the first three forest associations share one characteristic which distinguishes them from the other examples, namely, the phanerophytes are represented in higher proportion than any other class or at least rank equal first in position together with hemicryptophytes (to within 1%). Geographically these three examples are from the eastern United States south of the territories to which the remaining types belong. In the Sugar Maple, Aspen and Poplar associations it is the hemicryptophytes that predominate in the spectrum and phanerophytes take second place with approximately half the values attained by the former. The two associations of the more atlantic regions, Long Island oak forest and Laurentian maple forest, show clearly greater representation of chamaephytes (10%) while the associations of the continental interior, Michigan and Alberta show greater importance of therophytes than other regions. Perhaps contrary to expectation, cryptophytes, which Raunkiaer believed to be best protected from adverse conditions, are found in highest proportion in the two southern associations which on any criterion experience the least severe winter climates.

Table 7 Comparison of certain American deciduous forest associations

	No. of Species	Ph	Ch	H	Cr	Th
Cove hardwoods mixed mesophytic climax, Great Smoky Mountains, Cain, 1945	113	36.3	4.4	30.1	25.8	3.4
Mixed mesophytic climax, Cincinnati area. Withrow, 1932	127	33.6	3.9	34.4	23.4	3.9
Quercetum montanae, Long Island, New York, Cain, 1936	92	34.8	10.9	32.6	20.6	1.1
Maple association, Laurentian region. Dansereau, 1943	346	17.0	10.0	56.0	15.0	2.0
Aspen association, northern Lower Michigan. Gates, 1930	310	22.9	3.9	47.1	16.1	10.3
Poplar association, central Alberta. Moss, 1932	170	25.8	1.8	48.2	17.1	7.0

The possibilities for mapping vegetation on this basis are increased by the use of computers for data-processing. It offers the prospect of a new kind of vegetation map in which quantitative values could play a part. It would almost certainly confirm Raunkiaer's thesis that life-form composition exhibits a strong correlation with climate although not necessarily in a causal relationship as he supposed. Indeed the opportunity for refining this correlation exists since many independent variables of climate might be individually tested for coincidence with life-form representation.

Cain (1950) discussed an alternative means of using life-form spectra by relating places at which the proportion of a given element, e.g. chamaephytes, attains the same numerical value. The chamaephyte value of 27% occurs at an elevation of 1000 m in the Clova district of Scotland which has a relatively oceanic climate while the same percentage is not attained in the more continental climate of the Alps until elevations of 2500-2700 m are reached. The proportion of chamaephytes tends to increase with altitude and at 3600 m in the Alps the value of 30% is reached. Incidentally, this does not appear to be true in the southern Appalachians (Table 6). There is a hazard in relying on proportional values, however, in that an increase may result either from an addition to the number of species with that particular life-form or from a decrease in the representation of other life-forms. Moreover the actual number of all species comprising the vegetation type to be compared in two areas usually differs so that the same number of chamaephytes, possibly exactly the same species, would register slightly different percentages in the two cases. It seems unrealistic, therefore, to place significance on the precise figure but better to consider the balance in representation and importance revealed by order of ranking.

Desert climates and vegetation types

The belief that particular climates engender corresponding types of vegetation regardless of geographical position can be refuted convincingly by the use of life-form spectra. A comparison of the plant composition of several desert areas climatically classified as "hot deserts" (Table 8) reveals as many differences as there are similarities. The points in which they resemble one another are the pre-eminence of therophytes (though the percentage values are varied) and the low proportion of geophytes. It is worth noticing that stem succulents such as cacti form only a small but constant proportion of the desert flora. The remaining classes, including all the woody plants which exert considerable effect on the general physiognomy and structure of vegetation, are very variable in their representation. Nano-phanerophytes (shrubs less than 2 m high) are well represented in the California desert (21%) and in Australia (23%) but the latter has even taller woody vegetation in abundance as indicated by the 19% micro-plus meso-phanerophytes (2-30 m high) compared with only 2% in Death Valley. The North African desert has extremely few phanero-phytes of either sub-class (3% in total) though dwarf shrubs of low stature (included among chamaephytes) partly compensate this deficiency. There is no doubt that these samples would be classified as different vegetation types by all of the methods discussed in this chapter.

Table 8   Comparison of desert life-form spectra (extract from Cain, 1950)

	Stem succulents	Meso-phanero-phytes	Nano-phanero-phytes	Chamae-phytes	Hemi-crypto-phytes	Geo-phytes	Thero-phytes
Death Valley, California	3%	2%	21%	7%	18%	2%	42%
Ghardaya, Algeria	3%	0	3%	16%	20%	3%	58%
Ooldea, South Australia	4%	19%	23%	14%	4%	1%	35%

The dogma of vegetational convergence under climatic influence seems to have outlived its usefulness as a guiding concept. While there is a very sensitive correlation between the physiognomy and composition of vegetation and climatic parameters within each continent, differences in geomorphic history and in the evolutionary development of living things in the separate realms have produced different solutions to the problems of populating earth's varied environments.

SOURCES OF REFERENCE

Cain, S.A. (1945) The biological spectrum of the flora of the Great Smoky Mountains National Park. *Butler University Botanical Studies.* Vol. VII, 1-14.

Cain, S.A. (1950) Life Forms and Phytoclimate. *The Botanical Review* 16, 1-32.

Dansereau, P. (1957) *Biogeography: An Ecological Perspective.* Ronald Press, New York.

Dansereau, P. and Arros J. (1959) Essais d'application de la dimension structurale en phytosociologie. *Vegetatio* IX, pp. 48-99.

Fosberg, F.R. (1967) A classification of vegetation for general purposes. pp. 73-120 in *International Biological Programme Handbook No. 4* edited by G.F. Peterken. Blackwell, Oxford.

Küchler A.W. (1964) Potential natural vegetation of the conterminous United States. *American Geographical Society Special Publication No. 36* New York.

Miller, A.A. (1957) *Climatology.* Methuen, London.

Polunin, N. (1960) *Introduction to Plant Geography.* Longmans, London. (Includes as frontispiece a full colour world vegetation map by A.W. Küchler.)

CHAPTER 8

Migration and dispersal

Seasonal and periodic migration

The spreading of plants and animals into new territory is their principal means of adjustment (as species) to major changes in their physical and climatic environment, especially changes that are gradual in effect and of long duration. Such movements of populations are usually called migrations though the same term is used also for the regular periodic movements of birds and mammals that relate to seasonal change. Migration, therefore, includes both short-term and long-term translocation, both temporary and permanent, but the latter is more significant in biogeography.

Among animal species every kind of migration is represented from the most ephemeral to the most enduring and a brief summary of some examples will help point out their several characteristics. *Seasonal migration* is temporary short-term movement of regular periodicity which occurs, for instance, when swallows leave their breeding areas in Europe at the end of the northern summer to spend the other half of the year in southern Africa. Although the seasonal migrations of birds and of marine mammals, such as whales and seals, are the most spectacular because of the great distances covered, very many terrestrial mammals, particularly the larger herbivores such as elk (moose), reindeer (caribou), and impala, regularly complete seasonal migrations over still very considerable distances. The extent of territory covered in short-term and seasonal movements is exemplified by the map of wolf trails (Fig. 8.1).

The characteristics of seasonal migration, resulting from the mobility of animal species, are that movements are repeated annually in both directions and thus many times within the life-time of the individual.

Temporary migration of a different kind occurs in some animal species in which the migratory journey is made only once in each direction during the life of the individual, as in the North Atlantic Eel, or on more than one occasion but at variable intervals of more than a

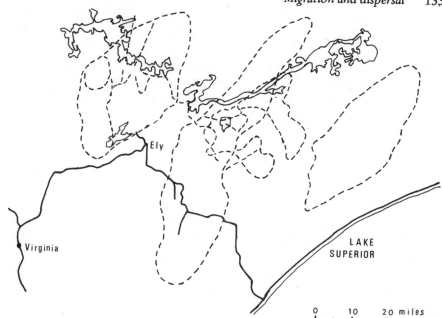

8.1 Wolf trails in north-east Minnesota and adjoining Ontario (after Olson 1938), showing larger lakes in the area. The routes are followed by hunting packs on circuits of 2-3 weeks' duration. Lake crossings only passable in winter.

year, as in the Atlantic Salmon. In such cases the migrations are not related to seasonal change in climate or availability of food: they are associated with specific stages in the life-cycle often accompanied by changes in body form. In 1921 Johannes Schmidt discovered that the North Atlantic Eel (*Anguila anguila*) breeds only in the Sargasso Sea, south-east of the Bermudas, and that only here are the youngest larval forms found. With the aid of the North Atlantic Drift these small fish migrate to European coasts from Iceland to Portugal, taking 2½-3 years to accomplish this journey. Continued migration from marine to fresh water takes them upriver where they remain for between seven and nineteen years before retracing the entire journey to the breeding-grounds of the Sargasso Sea.

The Atlantic Salmon (*Salmo salar*) differs in having freshwater breeding-grounds. After a variable period of 1-4 years from birth in the swift-flowing rivers of western Eurasia and eastern North America, the young fish or parr migrate downstream. They spend the next 1-4 years (exceptionally 5 years) as a sea-fish before returning to the rivers for another variable period in which breeding takes place. Some individuals return to the sea and of these a proportion make subsequent spawning migrations up the rivers.

These examples are distinct from seasonal migrations because the two-way journeys are not completed within a year. The individual may complete the return migration only once in some cases and the interval between marine to freshwater transitions is indefinite and usually lasts several years.

However, like seasonal migrations, these are two-way movements and they are annual occurrences for the species as a whole, even if not for the individual. They are therefore classed as *periodic* migrations. Only when the offspring of the migrant generation do not return to the original territory of their parents is there the possibility of permanent migration.

Permanent migration

By definition, permanent migration involves encroachment into an area previously unoccupied by the species: it need not imply abandonment of the previously existing range. Migration in this sense thus involves extension of range.

It applies equally to plants and animals, including species which migrate seasonally if, for the sake of illustration, the area occupied by breeding populations or frequented as pasture grounds were extended gradually year by year. In the case of animals that are normally non-migratory, e.g. badgers, beavers, etc, the extension of range is a less obvious process which is effected through the young adults of a new generation. If when moving from the parent colony to establish their own territory (a new badger sett or a new beaver dam) these individuals move beyond the fringe of the area in which all previous colonies were sited they extend the frontiers of that species. It follows that the advance of a species may normally be a slow process, observable perhaps only over several decades and measurable mile by mile, being determined by the slow progression of generations as one succeeds another in establishing territorial claims.

A northward migration of the opossum in Wisconsin (Fig. 8.2) has been reported (Long and Copes, 1968) and due to the existence of earlier records its history within the state is known fairly accurately over the past century and its rate of advance has been calculated. Up to 1872 the opossum was not known to occur north of Madison and in the next half century little change was noticed. In 1920 a specimen was taken at Beaver Dam, Dodge County, and after this date the species must have spread rapidly. By 1958 it had reached Oneida County, 160 miles to the north, and in 1967 it was reported both to the west and to the east of this line in Clark Co. and Kewaunee Co. respectively, both of which are over 100 miles from Beaver Dam. It is therefore certain that permanent migration of the opossum can take place at a rate which extends the territory of the species by 2-4 miles per year. This situation is very much akin to that in plants, which on first consideration do not appear to possess that fundamental requirement for migration — mobility.

Mobility does exist, however, in the dispersal stage of the plant

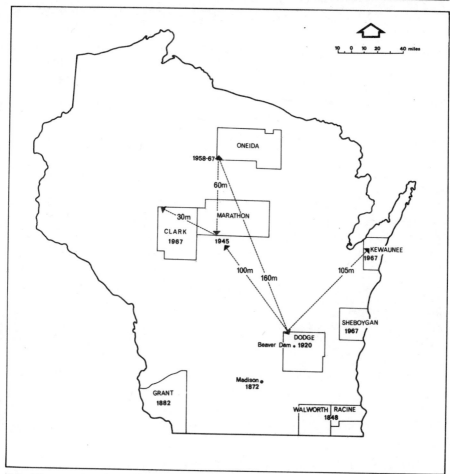

8.2   The spread of Opossum in Wisconsin. The dates given record the first sighting
in each of the counties named. Data from Long & Copes, 1968.

life-history when the reproductive body or propagule is transported by
such means as wind, water, birds, fruit-eating bats, etc. The distance
and direction of transport is haphazard and depends greatly on
fortuitous circumstances at the time when the propagules are released.
These agencies are a normal part of the dispersal process in plants and
at the same time are the vehicle of plant migration. While in some cases
a detached vegetative organ of the plant body, e.g. a tuberous organ
(turion) or a daughter rosette in floating water plants, may provide the
means of dispersal, ordinarily it will be the seed or a detached fruit.

The possibility of migration in plants therefore depends upon the
fulfilment of three conditions. The existing mature plants must produce
viable propagules, e.g. fully developed ripe seed, and this condition is
not always met, particularly in places near to the edge of the species
range where climatic limitation may operate. Secondly, dispersal must
be effected into territory outside that occupied by the parent

population. Thirdly, the propagule must successfully establish itself, i.e. take root and commence independent growth. An area of varied country can be considered occupied by a species when populations are present at localities within that area at which suitable habitats occur. Migration involves the extension of this area by the arrival and establishment of plants in localities beyond the previous limit of occupied habitats. The colonization of unoccupied habitats at places *within* the original territory does not qualify as migration and these two terms should be distinguished.

Migration thus defined may express itself geographically in several ways. It may produce a simple expansion of the species range or distribution; alternatively expansion in one direction may be accompanied by withdrawal in another quarter so that the distribution as a whole shifts latitudinally. Cases in which the distribution contracts do not involve occupation of new territory nor necessarily any movement of individuals or populations if death removes them. Strictly speaking, therefore, no migration takes place but purely as a phenomenon of distributional change it is convenient to refer to a contracting boundary as "retreating migration" (Gleason and Cronquist) or migrational retreat.

It is clear from the above discussion that plant migration is entirely dependent upon the function of reproduction and that in animals, too, locomotion does not supplant the role of reproduction. Increase in population is essential for it to occupy a larger area. In a dynamic situation the distance covered by migrational advance will be a function of the number of generations passed, taking into account also the fecundity of the species, the number of progeny surviving to maturity and the length of the life-cycle. The mechanism of migration can be best appreciated by considering a hypothetical situation in greater detail.

Good's theory of plant migration

Consider a tree species which, we will suppose, has a temperature tolerance limited by minima of $0°C$ and by maxima of $30°C$. Imagine that there is a gradual climatic change involving a lowering of temperatures within the area of the species and that the temperature limits of $0°C$ and $30°C$ move southward. The result is that the area occupied by the species and the area of appropriate climatic conditions do not now coincide completely. The climate has been displaced southwards of that area leaving a zone in the north where the existing trees find themselves in a climate no longer favourable, and creating in the south a zone beyond the boundary of existing trees which has become climatically suitable for their growth. Near the northern margin of the plant's range the growing trees will be adversely affected by the deterioration in climate but this does not necessarily cause their death. More probably the trees there may fail to produce ripe seed as a result of the inadequate summer warmth or if seed production is unaffected the too severe winter cold may prevent the survival of seedlings. Consequently, although the standing trees may survive, their reproduction has ended, no succeeding generation is being established to replace them and on their eventual death the species will cease to exist in this zone. This sequence of events represents the mechanism of migrational retreat.

At the southern edge of the original distribution, when this coincided with the isotherm of 30°C, the plants were prevented from spreading further south by temperatures in excess of the maximum in which their seedlings could survive. This means that a proportion of seed produced by the trees here was dispersed (to the south) into an area where seedlings could not establish themselves.

As a consequence of the climatic change the limiting maximum temperature of 30°C no longer occurs in localities at the southern boundary of the species and the seeds dispersed from plants growing here now succeed in producing new individuals which in their turn reach maturity. From their seed a second generation is established including a proportion dispersed again southward and extending further into the zone newly available to the species through the removal of the climatic restraint. Thus at the southern margin, at the advancing front of migration, successive waves of seedlings progress further into new territory with each generation. It is worth noting that no "response" is necessarily evoked in the plant by the changed climatic situation. Its functions continue as before with the difference that while previously its southward dispersed seed came to nothing, in the new situation it is viable.

We have assumed throughout that from each individual in the population seed is dispersed radially to varying distances, the number of seeds diminishing with increasing distance from the parent. For every species there will be a certain maximal distance to which seed is dispersed. This distance together with the average length of life-cycle from germination to seed production will determine the rate of migration. Other things being equal, an annual herb has opportunity to advance every year but a tree that produces no seed for its first 10-15 years may advance only at intervals of this duration. The infinite combinations of dispersal range and reproductive age make it clear that the potential rate of migration is different for every species. The actual rate of migration will be further modified in every case by the degree of success in establishing seedlings in competition with plants of other species. It should not be forgotten that generally migration does not take place in unpopulated virgin territory: almost always some other species are already in occupation of the area and the rate of progress of an immigrating species is conditioned by its ability to compete with the plants of the existing vegetation. The actual rate of migration is also modified by other external factors, such as the spatial distribution (frequency) and areal extent of suitable habitats in a particular landscape, and by dependence on specific animals as agents of seed transport. In combination the operation of all these factors is so various that to predict a migration rate for any species would be highly unreliable. Knowledge of migration must be acquired through careful study of the dispersal-power of each species and by measuring migration time and distance in the case of past population movements whose occurrence is recorded either by historical documents or by fossil evidence.

Effects of topography on plant migration

In the hypothetical example described above it was assumed that migration takes place in an area of uniform topography. In real situations, varied topography complicates the way in which populations move.

If the area initially occupied by the species is a plain flanked on both sides by mountain ranges which diverge it is likely that the additional area made available by the climatic shift would enlarge more rapidly than the plant could spread into it. This suggests one reason why plant migration may lag behind a climatic change in its favour. Of course, species with inherently slow rates of spread may do so even in the absence of such considerations.

Wulff has drawn attention to another effect of mountain ranges that can be most important for plant migration. In all cases of widespread climatic change the influence extends not only in latitude and longitude over the lowlands but also displaces the climatic zones on mountains either upward or downward. This vertical climatic shift provides opportunity for altitudinal migration. A given climatic zone may be displaced only 2,000-3,000 ft upwards on a mountain during a period of increasing warmth, while on the plains the equivalent climatic conditions may have shifted 200-300 miles to the north. Where great distances are involved there is a risk that some slow-spreading species will not migrate fast enough and may be overtaken by unfavourable conditions. However, in mountain areas species may spread to higher elevations where appropriate climatic conditions continue to exist nearby. Because migrational adjustment can be made over small distances in this case, mountains often serve as areas of survival (refugia) for species which may have lost their ground in lowland situations through their inability to keep pace with rapid environmental

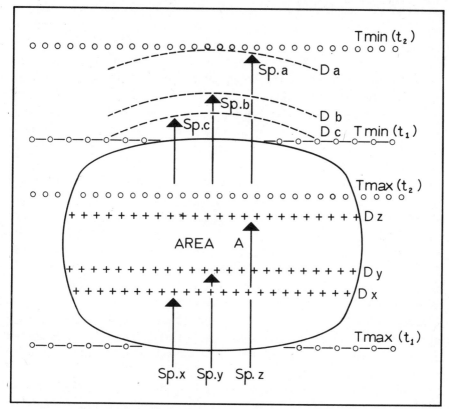

8.3 Schematic model of plant migration under the influence of a climatic change. A latitudinal shift of isotherms (T) is assumed. After an interval of time (t_1-t_2) the distances (D) through which several species (Sp) have migrated is represented by arrows.

change. Of course such refugia may become centres of dispersal if suitable conditions return subsequently to the adjacent lowlands.

Floristic consequences of migration

Now reconsider the hypothetical situation, discussed on page 136, of an area which experiences a general climatic change, but this time imagine that the area (A) in Fig. 8.3 is populated by a large number of plant species comprising a forest. The mean summer temperatures prevailing at the southern and northern boundaries of the area before the change in climate are denoted as T max. (t_1) and T min. (t_1) respecively. The climatic shift of this temperature zone results in the northward movement of these isotherms to the positions marked T max. $(t_2$ and T min. (t_2) respectively. Many species of area (A) now find themselves able to extend their distribution northwards which they do at various rates determined by dispersal range, competitive success etc. At time (t2) species a,b,c, have migrated different distances to the north and have not yet reached their potential limits. It may be that, in some cases, their southern boundaries are at the same time in the process of migrational retreat if the summer temperatures in the southern part of area (A) now exceed their limits of tolerance. Meanwhile, species x,y and z, having higher temperature requirements and formerly restricted to an area to the south of (A) by the limiting temperature T max., are now migrating into area (A), adding new diversity to its forest. Again, their rates of progress are individually different and the distances (D) to which they have penetrated at time (t2) are indicated in the diagram (Dx,Dy,Dz).

It is clear that on theoretical grounds the effect of a climatic change in an area of uniform vegetation will be to cause sorting or separation of species. The vegetation-type which existed under stable climatic conditions cannot migrate as a unit but is disrupted by the differential migration of its constituent species. New associations of species are produced to form new types of forest and, even if sufficient time elapses for migrational readjustment to cease, the reconstituted forests cannot precisely resemble their precursors.

Some of the plants originally in area (A) may have been limited to it by rainfall requirements and, providing this does not alter with the postulated change in temperature, such species would be unaffected. In reality, it is likely that a moisture-dependent species would actually contract in distribution if temperature rose since the resulting increase in evaporation would reduce available moisture and increase rate of water-loss. Of the migrating species, some would remain in occupation of the whole of area (A) while extending northwards but others, sensitive also to excessive summer warmth, would retreat from the southern part of area (A).

The homogenous forest would therefore never be reconstituted in its original form. The effect of major climatic change is to cause re-distribution of species and the intermixing of floral elements from different regions. The flora of a large region is made up of species from different areas of origin which migrated at different historical phases in the long sequence of its climatic history and vegetational development.

Dispersal in plants

Dispersal of the individual is the mechanism of migration and even an approximate idea of the rates at which tree species migrate can only be gained from precise observation and experiment. A study of seed dispersal and survival in Engelmann Spruce (*Picea engelmannii*) reported by Alexander (1969) illustrates the factors involved and allows some estimate of the migration rate for this species to be made. This tree is regarded in forestry as a moderate seed producer (presumably in

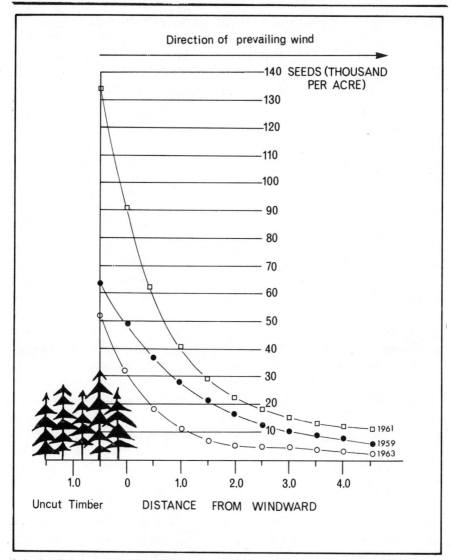

8.4 Seed dispersal in Engelmann Spruce (*Picea engelmannii*). The seedfall within the forest (at left) and at various distances (measured in chains) from the forest edge is shown for three years of significant seed production. Redrawn, from Alexander 1969.

comparison with other softwoods) but the light seeds are well dispersed by air currents. Even so, 50% of the seed dispersed falls within 1.5 chains (30 m) of the forest edge. At 4.5 chains (90 m) the seedfall amounts to only 10% of that recorded within the forest and according to Roe (1967) less than 5% of the seed production of a stand is dispersed as far as 10 chains (200 m) into open ground. These figures refer to seedfall measured to the leeward side of a forest edge, i.e. dispersal aided by prevailing winds (Fig. 8.4). The actual amount of seed this represents varies greatly from year to year with the seed crop produced by the parent trees. This is a very common observation in silviculture and applies equally to other species of trees. A good crop of seed in Engelmann Spruce produces in excess of 100,000 viable seeds per acre within the forest and exceptionally may attain ten times that value (a "bumper" crop). However, a good seed crop occurs only once in 5 years on average and a "bumper" crop once in 12-26 years at irregular intervals.

At 10 chains (200 m) from the forest edge, therefore, the seedfall could reach 50,000 per acre in exceptional years but in most years it would be fewer than 1,000 per acre. However, it is not enough to know the number of seeds reaching the ground: the number of successful seedlings produced is more relevant. Alexander reports that establishment rates of 1,000 seedlings per acre over a ten-year period were observed only within 1 chain (20 m) of the parent tree. In open ground further from the forest edge rarely more than 250 seedlings per acre had become established in ten years from a total seedfall of about 50,000 per acre over the same period of time. Considering the mortality that is bound to occur from natural causes during the next ten years of growth, this density of seedlings is regarded as marginal for forest regeneration. It can be questioned whether such thinly dispersed individuals can survive in the case of spruce and similar tree species which require the shade and humidity created by the existing forest to grow from seedling stage to maturity. Alexander concluded in this particular case that seedlings of Engelmann Spruce require protection from intense solar radiation, high and low extremes of temperature, frost-heaving of soil and the effect of drying winds. These conditions are found only within 20-30 m of the forest edge so it is here that seedling establishment is reliable, though effective dispersal of seed reaches greater distances.

From the evidence of this investigation we can try to estimate the rate at which Engelmann Spruce might spread into new ground at an advancing forest front. If 1 chain (20 m) is reliably seeded in every year and a tree becomes reproductive in 10 years, then the forest could advance this distance per decade. Alternatively, regarding the "bumper" seed years as the really effective events in perpetuating the species and taking 20 years as the average interval between such events, seedling establishment might occur in leaps of 10 chains (200 m) on these occasions. This represents an average rate of forest advance (migration) of 50 chains (1 km) per century. Of course, changes in the incidence of heavy seed years or in any of the other variables would modify this figure. However, the experimental conditions in this example, involving the artificial clearance of a patch of forest to create an opening, do not closely simulate the natural situation in which Engelmann Spruce and other climax forest trees could be expected to extend their range. As noted already, the seedlings of this tree demand the shade and humidity of a forest environment for their success and this could well be provided by a forest of other species. Thus the natural migration of spruce would be expected to occur, not across an unforested landscape, but by invasion and spread through a forest of some other species such as pine. The rate of migration achieved by spruce under these conditions would be greater due to the higher establishment and survival rates of seedlings from low density seedfall at the limiting distance of its dispersal range. The distance of seed dispersal would not be increased of course, but seedling growth from the 5% of seed dispersed as far as 10 chains (200 m) from parent trees would be more probable even in normal seed crop years. At the most optimistic estimate the species might advance this distance annually and therefore its migration rate could conceivably be 20 kilometres per century.

The value of using data from modern forestry experiments is that it provides quantitative information on the distances to which tree seeds are dispersed by natural means. This is something which cannot have changed appreciably during the history of present species. It is also helpful to learn from experiment what factors influence seed dispersal. Direction and velocity of prevailing winds can alter the distance to which seed is dispersed in different years. The annual variations in seed crop do not affect the *distance* to which seed is dispersed although the *density* of seedfall per unit area and therefore the chances of successful establishment does vary as a consequence at all distances reached from parent tree (Fig. 8.4).

Dispersal in beetles

In some species of ground-beetles (Carabidae) the membranous wings of all individuals are reduced in size and as a result they are unable to fly. Other species have entirely full-winged populations. In a third group there are species containing some individuals of short-winged form (brachypterous) and some of long-winged form (macropterous). Those species exhibiting two forms (dimorphic) provide interesting material for studying the dispersal power of flighted individuals compared with those whose only means of locomotion is running. Lindroth has observed that within the geographic range of dimorphic species, long-winged and short-winged individuals are not uniformly distributed: towards the periphery of the geographic range the fully winged form outnumbers the reduced wing. He concludes that long and short-winged individuals must have different powers of dispersal and that the former act as parachutists, taking a more active role in advancing the frontiers of the territory occupied by the species.

A natural opportunity to observe the process of dispersal during immigration of the species into new territory is described by den Boer (1970). In the Netherlands land reclamation from the Zuider Zee is in progress and within the area of distribution of a dimorphic species (*Trechus obtusus*) a virgin territory was created when the polder of East Flevoland ran dry in 1957. After a period of seven years a population of this beetle was discovered on the new land area now exposed at a distance of 1,000 m from the old land and separated from it by a freshwater lake 700 m wide and 5 km long. An exceptionally high proportion of the new population (95%) consisted of fully winged individuals. In fact only four short-winged individuals were discovered in the sample and these are thought to be the progeny of already fertilized females which were themselves fully winged and capable of flight. Thus long-winged beetles are able to cross barriers of about 1 km and there is a strong chance of such dispersal occurring within a short time. This is confirmed by the fact that in the seven years available for the immigration of beetle species into the polder, 45 species arrived, of which 27 were fully winged types and 17 were dimorphic species represented chiefly by long-winged individuals. The conclusion drawn by den Boer is that for totally flightless species "it seems almost impossible to reach this site within this space of time". A single individual represented the only flightless species found (*Carabus monilis*) and as this is a large beetle (24-28 mm) whose habitat is river banks it is not very surprising that it reached the polder island.

Den Boer's observations revealed more rapid migration in *Bembidion varium*. This beetle of riparian habitats was understandably absent from the dune area north of The Hague until an artificial lake was created as a water-supply reservoir for the city. It then appeared within six months after the flooding of the reservoir although the nearest localities at which it was previously known are situated 25-30 km to the south-west. The prevailing south-west winds therefore must be effective in dispersing the beetles over such comparatively large distances.

Most carabids are considered to be weak fliers, having only slight directional control of their flight which is therefore strongly influenced by the wind. However, the fact that they are able to keep themselves air-borne puts them at the disposal of air-currents, including turbulent and thermal flow. Indeed, generally speaking it is the small insects and the weak fliers that are carried helplessly from one country to another and from these much of the insect population of remote areas is derived (Oldroyd). From remarks of this kind it would appear that flighted beetles, by virtue of their active buoyancy in air, outstrip plant dispersal units of comparable size and weight, e.g. "winged" seeds such as pine. They are probably equalled in their dispersal range only by the minute powder seeds and pappus-borne ("parachute") seeds. Of course we must recognise that in the case of *Bembidion varium* the individuals have not covered the full distance of 25-30 km without touching ground as a plant seed probably must do when released from a cone on the high branchlets of a tree. The beetle may be transported over comparatively short distances while flying on any single occasion and the dispersal range quoted is the aggregate distance drifted during numerous flights. However, the evidence of one or two examples should not imply that wide dispersal power is generally found in beetles of all kinds. Not all long-winged beetles are capable of flight, for a greater or lesser proportion of individuals (depending on the species) has undeveloped flight muscles, as Lindroth has pointed out. Moreover the habit of flight is seemingly characteristic of species occupying unstable habitats (of which riparian situations are an example) where the resulting scattered deployment of the population ensures its continuing survival.

In beetle species of stable habitats, e.g. forest, dispersal may be of a much lower order, contrary to expectation based on the presumed flying powers of full-winged individuals and the mobility of both types on the ground. *Pterostichus strenuus* is a dimorphic ground beetle that inhabits deciduous woods. Observations by den Boer from trapping experiments show that flightless individuals have very small chance of covering a distance greater than 100 m. Some were trapped in damp heathland at distances up to 40 m from the nearest wood but even a ditch or stream would form an insurmountable barrier to their dispersal. At distances of more than 100 m from deciduous woodland only the full-winged individuals were encountered. Other woodland species whose populations include a proportion of full-winged individuals apparently disperse even less since specimens were hardly ever caught outside their preferred habitat, e.g. *Pterostichus diligens, Bembidion lampros.*

The dispersal power of flightless species, in which all individuals have reduced wings can be judged, according to den Boer, from data he obtained on the largest of such species in the Netherlands, *Carabus problematicus* (21-28 mm), which is a "good runner". It is a species found in dry woods, including pinewoods, and open heath. Individuals occur commonly up to 500 m from their point of origin and exceptionally at distances up to 1 km but the chances of their crossing gaps of 4 km between isolated woods are negligible.

This evidence shows that carabid beetles, which are probably about the most active and mobile of all invertebrate animals, are capable of dispersing over distances of several hundred metres from established populations; and in certain "opportunist" species dispersal may be measured in tens of kilometres in directions favoured by wind conditions. However, in trying to estimate potential rates of migration, it cannot be safely assumed that such distances can be covered annually: indeed it is most unlikely. The dispersed individuals must be present in sufficient density for breeding encounters to occur and at the extreme limit of dispersal range this will not be so. Exceptionally, already fertilized females may take part in dispersal but otherwise reproduction of the dispersed individuals must take place to provide a new stock before further progress in migration can occur. As dispersal of individuals is assumed to be directionally random only a minor proportion of each generation will be moving in the "right direction", i.e. into territory not already occupied by the species. Only this fraction will contribute to the migrational advance (cf. our hypothesis of plant migration). So far as evidence is available, it suggests that any realistic estimate of the migration rates of beetles should be based upon the dispersal distance reached by abundant individuals, from which new breeding populations can be established to serve as new dispersal centres in succeeding years.

Conclusions

The dispersal distances for beetles appear to be not very different from the distances that plant dispersal organs are transported by wind. Even the large distances reported for "opportunist" beetles have their equivalents among flowering plants in those weeds and pioneer colonists of temporary habitats which have powder seeds or are aided by a parachute pappus of hairs e.g. Fireweed (*Chamaenerion angustifolium*).

A potent complicating factor in translating possible rates into actual migration is the frequency of habitats suitable to the requirements of the species in question. The simplest case is where a virtually continuous habitat, such as unbroken forest, exists in the path of plant or animal species adapted to this environment. On the other hand, the progress of species of disturbed habitats, such as riverbanks, within generally forested terrain would be less predictable.

"It is quite evident that, when man does not contribute to the dispersal, it goes on very slowly . . . It is clear that no single indigenous plant . . . can be expected to have spread around the globe, and hardly even across America, during the approximately 16,000 years of postglacial time. Halfway across the North American continent, that is to say, some 2,500 kilometres, is apparently a more normal distance for plants to cover in 16,000 years. Turning to the conditions in Europe one is inclined to consider this figure too high, rather than too low. Misconception of this rate of progress has . . . been one of the most serious obstacles in the way of arriving at an approximately correct idea of the processes that have led up to the present distributional conditions within the boreal belt." (Eric Hultén 1937.)

SOURCES OF REFERENCE

Alexander, R.A. (1969) Seedfall and establishment of Engelmann Spruce in clearcut openings. *U.S.D.A. Forest Service Research Paper* RM-53 8 pp.

Boer, P.J. den (1970) On the significance of dispersal power for populations of Carabid beetles. *Oecologia* 4, pp. 1-28. Berlin.

Good, R. (1953) *The Geography of the Flowering Plants.* Longmans, London.

Hultén, E. (1937) *Outline of the History of Arctic and Boreal Biota during the Quaternary Period.* Stockholm

Lindroth, C.H. (1949) Die Fennoskandischen Carabidae III. *Göteborgs kgl Vetensk. Handl.* B4 (3), pp. 1-911.

Long, C.A. and Copes, F.A. (1968) Note on the rate of dispersion of the opossum in Wisconsin. *American Midland Naturalist* 80 (pt. I) pp. 283-4.

Oldroyd, H. (1966) *Insects and their World.* British Museum (Natural History), London.

Olson, S.F. (1938) Organization and range of the (wolf) pack. *Ecology,* 19, pp. 168-70.

Roe, A.L. (1967) Seed dispersal in a bumper spruce seed year. *U.S.D.A. Forest Service Research Paper* INT-39. 10 pp.

Wulff, E.V. (1943) *An Introduction to Historical Plant Geography.* Chronica Botanica, Waltham, Mass.

CHAPTER 9

Relicts and refugia

Under this title we shall discuss features of geographical distribution that suggest that a species, a group of species or a stand of vegetation is relict in a certain locality or region. There are other means of arriving at this conclusion when fossil evidence is available but as it is not always available, and for thousands of species probably never will be, the biogeographic evidence of distribution must provide the signs that tell of their previous history. In following these signs the preceding discussion of migration and of the conditions and distances which create barriers to migration will be useful if kept in mind.

Geographical relicts

If the area occupied by a species is progressively restricted, for example, by general climatic change or by competition from an invading species, the species will survive in those places scattered within its former territory where that particular change of climate has least effect or the competitor has least advantage. In certain localized situations the climatic effect may not exceed the tolerance limit of the species where appropriate protection (e.g. from excessive warmth) or enhancement (e.g. where warmth is a requirement) is afforded by the topography. Thus, steep-sided ravines with their prevalent shade do not experience the full warmth of more exposed situations. On the other hand, south-facing slopes protected perhaps by crags from the worst effects of cold winds may possess a microclimate considerably warmer than that of the surrounding landscape. Such situations are favoured by purely local circumstances and therefore they are likely to be separated, one from the next, by relatively large distances. The consequence of survival only in these special situations is that the distribution of the species would become fragmented, the originally continuous (and extensive) area having been broken up and reduced to a number of small, isolated areas. The condition thus produced is what we have formally recognised as disjunct and discontinuous distributions. The scale on which

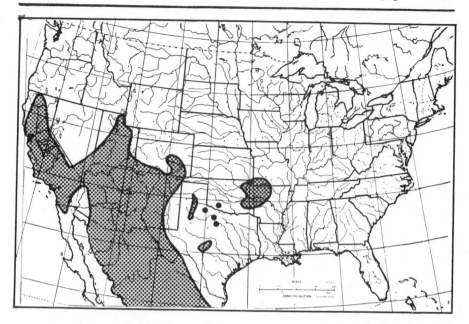

9.1a Distribution of the Brush Mouse (*Peromyscus boylei*) showing disjunct populations in Texas and in the Ozarks, far distant from the boundary of its main range in New Mexico. (After Blair, in Hubbs 1958.)

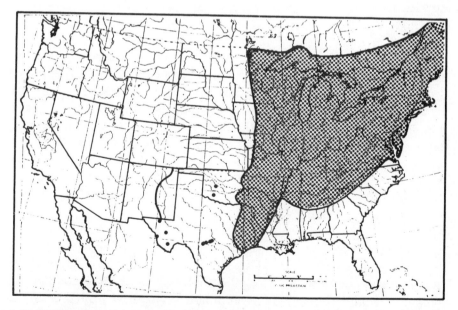

9.1b Distribution of Sugar Maple (*Acer saccharum*) showing relict occurrences in Texas and Oklahoma. Its main range is stippled. The eastern limit of *Acer grandidentatum* in the mountains of New Mexico is also shown. Adapted from Blair, in Hubbs 1958.

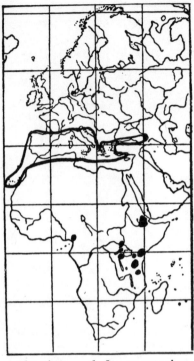

9.2 Disjunct distribution of the
mediterranean Tree Heath
(Erica arborea) in the mou-
ntains of Ethiopia, East
Africa and Cameroun.
After Rikli, 1943.

reduction and fragmentation of area occurs can vary between wide
extremes depending on how extensive the original area was and over
how much of it the species has been eliminated. Of course the process
may affect only the margin of the total (world) distribution of the
species so that it may still retain a large and continuous range over
territory where the adverse influence has not penetrated. Thus, various
forms of area can result from partial and complete contraction of a
species but the feature common to all of them is the existence of one or
many "islands" of occurrence at some comparatively great distance
from the residual centres of distribution, i.e. the largest territories still
occupied. The disjunct areas, whether they are individually extensive,
e.g. Brush Mouse in the Ozarks (Fig. 9.1a), or so small that they are
better described as stations or localities, e.g. Sugar Maple in Texas and
Oklahoma (Fig. 9.1b), are described as geographically relict. The places
in which relict populations survive are known as refugia because it is in
these locations that they have escaped adverse changes that have caused
deterioration of the surrounding environment.

The possibility that this kind of distribution might be an
expanding one and that the scattered pockets of occurrence represent
the pioneer populations of an advancing migration is not supported by
the available facts concerning the dispersal of individuals and the
progress of migration in modern populations that have been studied. As
an alternative explanation this has never been found acceptable by
those biologists best qualified to assess it. Thus, Stanley Cain wrote,
"An expanding area tends to have a relatively continuous boundary . . .

a contracting area tends to have a relatively discontinuous boundary. The reasons are that the expanding area has not reached effective barriers and the contracting area leaves behind relic colonies in local situations ... Advance colonies ... do not show any such widely disjunct penetration ..." Eric Hultén also endorses this view: "It is just the stations found outside the "compact area" that are likely to be the most valuable ones, which can give a clue as to how the development (of the distribution) has taken place." The "relict" interpretation of disjunct areas clearly implies that the dispersal distance of the species concerned is incapable of bridging the gap that separates surviving populations. Just how limited the dispersal range often is has been illustrated by examples in the last chapter. Even the species that inhabit temporary and unstable environments, whether insects or plants, do not migrate by crossing distances as large as those involved in most disjunct distributions. It could be argued that since their habitats are always a minority element in any landscape their distribution is never continuous. This is a matter of scale: in the sense that all available habitats within an area are occupied by populations of that species and these local populations have been established by dispersal flight of individuals (beetles) or wind transport of diaspores (plants) from one to another, this is geographically a continuous distribution. Even in widely dispersed species, such as weeds or the beetles of unstable habitats, what we do not find is the invasion of one newly available site within dispersal range of an established population while at the same time other sites within range remain uncolonized. The remarkable thing about the opportunist species is that wherever a transitory habitat appears they arrive (within a few weeks or months), providing that the sites are not beyond dispersal distance from already occupied territory. The point of this digression is to explore adequately the migrational potential and pattern of species with the greatest dispersal power to be sure that these are no exception to the interpretation of disjunct and discontinuous distributions as relict. It is clear from abundant examples described by Charles Elton — especially numerous in case of weeds and insects — that the advancing front of a migrating species is indeed a continuous one, as dated records and observations have surely proved.

In another way the relict distribution contrasts with an expanding one: typically a relict species does not occupy all of the scattered situations that would appear to satisfy its environmental requirements. Highly dispersed as these potential sites are, the actually occupied sites are hyper-dispersed, i.e. they seem to be a haphazard selection picked on the hit-or-miss principle. This is not in accord with the processes of dispersal but it is explicable in terms of local extinction of declining populations in some of the sites and local survival of others. Minor differences in the numerical strength of the isolated populations and in the vigour of competitive species at the various sites can easily account for the apparently capricious success or failure of a species once its distribution has become fragmented.

Bi-continental disjunction

The circumstances in which a species may become relict, the size and location of relict areas and the scale of disjunction in such distributions are best demonstrated by considering particular examples. Beginning with the more spectacular and large-scale illustrations, the crocodile was formerly ubiquitous in the rivers of Africa from the coast of Natal to the Nile delta and in the great lakes of the eastern highlands. Its amphibious habit makes its dependence on freshwater and its wide distribution readily understandable. However, north of the Niger and west of the Nile extends the vast waterless territory of the Sahara and Libyan Deserts, having for the most part no continuous or permanent surface-waters. It is therefore improbable but true that the crocodile occurs in the permanent pools fed by ground water in the Tibesti Mountains, approximately in the centre of the desert area and about 800 miles from both Nile and Niger! Clearly there is no question of natural migration of the species under present conditions nor is it the kind of animal that would be transported by man for domestic purposes! Further speculation is not necessary because also in the Tibesti massif prehistoric rock drawings depict unmistakably the forms of giraffe and elephant, other species perhaps even more incongruous in that place today than the crocodile. There can be no doubt that at some earlier time the climatic conditions of the Sahara were considerably less arid; it carried a steppe or savanna vegetation and had normal surface drainage systems (the geomorphic evidence for this still remains). At such a time the widespread distribution of the crocodile must have continued north and west of its present limits to include the Tibesti Mountains. The present population of crocodile there has perpetuated itself since that time despite the fact that the species has disappeared from the vast surrounding territories with the increasing aridity and the vanished surface-waters. It is incontestably a relict population.

The mountains of East Africa provide other instances of relict distribution no less surprising in their implications. At high elevations in the mountains of Ethiopia, in the Ruwenzori range and on the eastern volcanic peaks occurs a tall shrub or small tree, the Tree Heath (*Erica arborea*). These separate mountain groups with large intervening distances comprise a disjunct distribution among themselves but in yet more splendid isolation Tree Heath occurs on Cameroun Mountain in West Africa and in the Canary Islands (Fig. 9.2). In addition to this highly fragmented distribution in Africa the plant has an extensive mediterranean range from Iberia to the Black Sea and is therefore

9.4   Illinois River at Starved Rock State Park near La Salle, Illinois.

regarded as a mediterranean element in the flora of Europe. It is not alone in possessing such distant disjunct areas: other European plants found on the East African mountains are the atlantic species, *Sibthorpia europea*, and the wide-temperate Wavy Hair-grass (*Deschampsia flexuosa*). Since solitary examples are always liable to attract criticism on the ground that they are "exceptional" it is worth mentioning that a number of north temperate invertebrate animals belonging to various groups, e.g. three species of spring-tails, three mites and one fly, also share this extremely disjunct pattern of distribution (Salt, 1954). To believe that these are innovations in the flora and fauna of the African mountains, their relatively recent arrival, and the means of dispersal that took them there, must be demonstrated. The more reasonable view is to accept that these are geographical relicts of formerly more continuous distributions. This viewpoint is supported by the existence of parallel cases in which the circumstances are better known and in which fossilized remains testify to the existence of linking populations in earlier times; but evidence of this nature will be the subject of separate discussion (Chapter 10). At the moment we will confine our attention to the circumstantial evidence of the distributions themselves.

Intra-zonal disjunction

There certainly are many more examples in which species are disjunct over lesser distances within a single biogeographic region and in the same climatic zone. In these cases there is no difficulty in recognising that the animals (or plants) formerly spread across the intervening terrain under more favourable conditions and that some kind of deterioration in climate has made the middle-ground no longer inhabitable for them. The Brush Mouse (*Peromyscus*) in south-western United States shows this kind of situation, so does the Sugar Maple west of the Mississippi (Fig. 9.1b). Several northern species of tree, including true boreal elements of the Canadian coniferous forest as well as species centred in the Great Lakes region, occur in isolated localities far to the south of their general limit of distribution. Northern White Cedar (*Thuja occidentalis*) has a wide distribution in the boreonemoral zone (Sjörs, 1963) around the Great Lakes and the St. Lawrence River and extends southward along the Appalachians at higher elevations. There are also scattered sites in Illinois and Indiana where this tree survives at low elevations, but only in certain protected habitats that provide a microclimate more akin to that of places further north. Typically these relict stations are in ravines such as French Canyon in the Starved Rock State Park, Illinois or on river bluffs of mainly northern aspect like those near La Salle (Fig. 9.3). The same situations occur along Sugar Creek in Turkey Run State Park, Indiana, where one

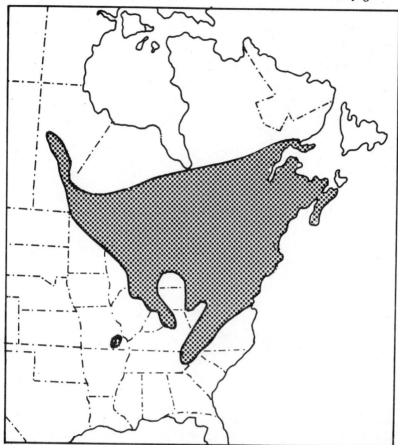

9.4   Distribution of Northern White Cedar (*Thuja occidentalis*) showing its disjunct occurrence in Missouri. (After Preston, 1950.)

can also find outposts of Canada Yew and Hemlock. As well as such localized occurrences in the northern states, Northern White Cedar is found 300 miles south-west in the Ozark Mountains of Missouri (Fig. 9.4). A comparable example is White Spruce (*Picea glauca*) which forms tall forest in the Black Hills of South Dakota, though isolated by 300 miles of prairie from the southern limit of its general distribution in Canada (Fig. 9.5). The Black Hills also have pine forests of Ponderosa and Lodgepole Pine, separated by 150 miles of treeless high plains from their main area of distribution in the Rockies. It is evident that the places where one species has outposts frequently harbour populations of other species similarly isolated from their main areas. It was this observation that first led to such localities being named refugia even before the mechanism involved was clearly understood. There can be no doubt that where species have distant outlying occurrences so far from their areas of continuous distribution and isolated from them by unsuitable environments, they stand as silent evidence that conditions at some time permitted their expansion to include these remote populations, which are therefore relict.

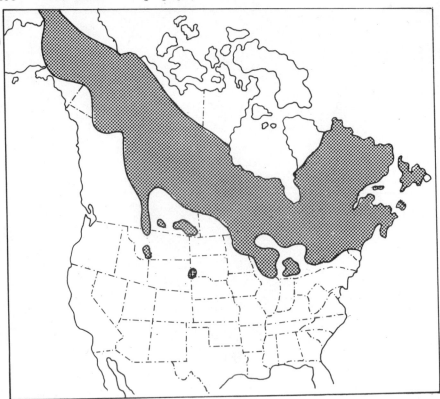

9.5 Distribution of White Spruce (*Picea glauca*) showing disjunct areas to the south of its main range, notably in the Black Hills, S. Dakota. (After Preston, 1950.)

In more extreme cases virtually the entire distribution consists of fragmentary areas and none of them are really extensive enough to be regarded as the main centre from which the others have become isolated: rather, they are residual parts from which the former existence of a wide distribution encompassing them all can be inferred. In Europe the large evergreen shrub, *Rhododendron ponticum*, illustrates perfectly this kind of situation. It has flourished in Britain and Ireland since it was introduced in 1763, as it has done in other maritime regions of Europe which experience mild winters. However, its largest territory of natural occurrence is in the pontic region from which it takes its name, i.e. north-eastern Asia Minor to the south of the Caucasus Mountains. It also occurs naturally in western Turkey and, remarkably, in three small hill areas in the atlantic fringe of Iberia and in the Lebanon Mountains (Fig. 9.6). Such a highly disrupted distribution is inexplicable except as geographically relict: only historical accident has prevented it from invading a wider territory outside these relict areas. The Cedars of the Mediterranean show similar signs of being residual from a period of far more extensive distribution (Fig. 9.7). Actually two separate species can be recognised in this case on the basis of minor differences. Atlantic Cedar is found at high elevations in the Atlas

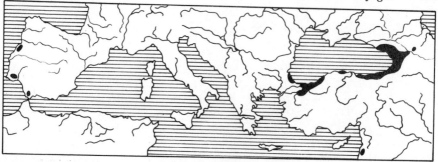

9.6 Distribution of *Rhododendron ponticum*, relict around the Black Sea, in Lebanon and in Iberia. (After Rikli, 1943.)

Mountains and Cedar of Lebanon in mountain districts in the eastern Mediterranean, especially the Taurus Mountains of Turkey. Since these are the only two species of their kind in the Mediterranean region and they are very closely related we can consider their combined (generic) distribution to be disjunct, but also both species considered separately have completely fragmented distributions which must therefore be relict. An almost exact parallel is seen in the case of Redwood and Giant Sequoia in California. These are the only living representatives of their kind and within their separate territories in the Coast Range and the Sierra Nevada respectively each has a discontinuous distribution, extremely localized in the case of Giant Sequoia (Fig. 9.8).

It should by now be clear that the pattern of geographical distribution can provide indisputable signs that part or even the whole of the area occupied by a plant or animal species is relict. This does not mean to say that the species concerned are destined for extinction or that they are outmoded in an evolutionary sense and unable to compete effectively. We have been discussing strictly geographical phenomena and in a majority of such cases the species has become relict, either locally or at large, as victim of changing conditions in the environment. In the rather common example of a plant species now restricted to separate mountain peaks and ranges, the plant has been forced to migrate to higher altitudes by a period of increased warmth or dryness which has made the lower elevations no longer tolerable. Thus the original continuity of its populations has been disrupted. However, it seems reasonable to suppose that if the climate should change back

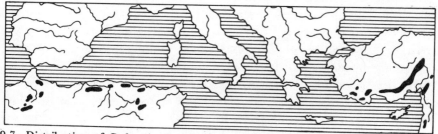

9.7 Distribution of Cedars in the Mediterranean region: *Cedrus atlantica* in North Africa and *C. libani* in the eastern Mediterranean. (After Rikli, 1943.)

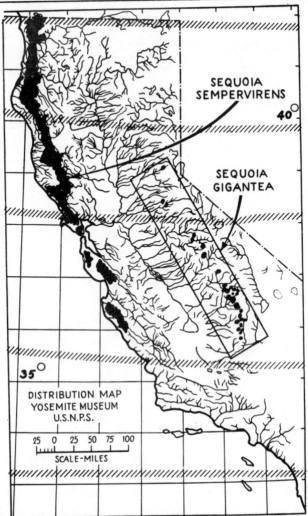

9.8 Distribution of Redwoods in California: *Sequoia sempervirens* in the Coast
Ranges (with minor discontinuities) and *S. gigantea* in the western Sierra
Nevada (highly fragmented distribution). From Bowers, 1965 by permission.

again downward altitudinal migration could take place and the species
could again establish itself over a broad territory. Indeed, there is no
need to rely upon hypothesis because stratigraphic and fossil sequences
have given us a very sound knowledge of climatic changes in many
areas. Climatic change has progressed through full cycle several times
during the most recent period of earth-history, the Pleistocene, giving
alternating glacial or pluvial phases and temperate or arid phases
(depending on geographical position).

The drastic effect of such major fluctuations has inevitably left its
mark on the areas occupied by plants and animals. Species with some
relict populations "stranded" in locations outside the main area of
present occurrence are not too difficult to understand against this
background but perhaps the examples of totally relict distribution

deserve further mention. Suppose that a plant or animal first becomes widespread during a period of extreme warmth and that it is then pressed southward into small coastal or peninsular areas during the ensuing cold phase of a climatic cycle. If climate then ameliorates but never again equals its first peak of warmth then the existence of mountain barriers, which under all conditions have a cooler climate than the surrounding regions, can prevent the species on its return migration from reaching suitable territory on the further side. Thus the situation frequently arises that relict species are contained by purely geographic barriers and therefore fail to occupy the more extensive areas climatically suitable for them. The interplay of biotic factors as well as changes in the physical environment make the real story of events more complex than this. The climatic cycle is repetitive but the areal pattern of living things is kaleidoscopic: when disrupted by catastrophic events the pattern never re-emerges in exactly the same form. We shall return to the question of what refugia are and what conditions they provide, but first there are other issues that must be clarified.

Evolutionary relicts

While species which are geographically relict in their total distribution may be quite able to compete in order to re-occupy a wider area when circumstances permit, there are among them some species which are relict in another sense. An evolutionary relict is a plant or animal that continues to exist after the extinction of most other members of its group. A relict species in this sense may or may not be confined to a very small geographical area. In such a complex subject it is rash to make generalizations that give the appearance of rules but nevertheless some guidelines must be given and the discussion of examples can at least provide a yard-stick by which other cases may be rationally assessed. It must be stressed that the only grounds for recognising evolutionary relicts are knowledge of the phylogenetic relationships of the organisms in question and fossil evidence of the ancient status of the groups to which they belong: geographic considerations are secondary here. However, it is often true that, in the living state, evolutionary relicts are endemic within very small areas. In many cases they were known as fossils long before it was discovered that they still have living representatives.

The flightless birds belonging to the ancient stock known as Ratites are considered primitive in several respects, including the lack of a skeletal keel to which flight muscles are attached. They appear to be descendants of birds which evolved before the faculty of flight became characteristic. The Dodo of Mauritius (another species on Reunion

Island) and the Moa of New Zealand were exterminated by man within the historic period but until man's intervention they were living relics in the evolutionary sense. In these cases the surviving populations were confined to very limited areas — in fact areas reduced to the logical end-point — a single island in each of the bird species mentioned.

The coelacanth fish, *Latimeria,* was first seen alive in 1936 when a specimen was found among a commercial catch taken off the coast of South Africa. The only related forms known were fossils of which the youngest became extinct 70 million years ago in the Upper Cretaceous period. *Latimeria* has its parallels in the plant world in the Maidenhair Tree and the Dawn Redwood. Both of these are endemic to China and within that country have very restricted distribution, the former being known only as a cultivated tree in the gardens of Buddhist temples which have ensured its survival. The Dawn Redwood (*Metasequoia*) was discovered in 1945 in the north-east Szechuan and Hupeh provinces, which seem to include the total range of the tree today. Both these plants have a long history of fossil ancestors. The name *Metasequoia* had already been given to some plant remains of Lower Pliocene age and other fossils extending back to the Cretaceous show the same characteristics. The Maidenhair Tree (*Ginkgo*) is distinct from all other living conifers and is placed in a class which is recognisable from Permian times in fossil representatives. Despite the extreme reduction of their areas trees such as the Cedars, Redwoods, and Ginkgo are quite able to tolerate climatic conditions in other parts of the world when grown in cultivation, as they now are for their ornamental value. It would seem that both their relict status and geographically relict distribution are consequences of waning success in natural competition with evolutionary products of more recent origin.

Refugia and the Pleistocene period

Now that we have seen the geographical context of species which are relicts of the evolutionary process, let us return to the geologically recent and the situation of species that are still widely distributed but have geographically relict outposts more or less remote from their main areas. Glaciation is only one of the influences that have disrupted plant and animal distribution and isolated relict populations but it will serve to illustrate the points at issue. During the Pleistocene period, ice-sheets extended southward from the north polar regions far into the heart of the northern continents where they have left a ubiquitous cover of till or boulder-clay and other deposits formed by the associated melt-waters. Wherever there is subdued relief, it is this superficial covering of glacial deposits that determines the shape of the modern landscape and is the parent material of the varied soils that have since developed. At the time of these glaciations all vegetational cover and animal life must have been displaced from the ice-covered territories. It can be assumed that plant and animal species migrated in reaction to the changes in climate that instigated glacial advance and in part resulted from it. If the progress of their migration kept pace with the expansion of the polar ice-sheets, we might expect that boreal forests and their biota persisted south of the glacial areas. Supposing we do not accept these assumptions, there is another possibility. If species generally failed to migrate their populations inhabiting territory in the path of the ice-sheets would gradually be eliminated while their populations in places not over-run by ice might survive *in situ*. If climate remained unaltered outside the glaciated area, vegetation there would be unaffected. The glacier snout, discharging vast rivers of meltwater, would come to its eventual halt among the debris of crushed and uprooted trees in the rich forests of the lower Mississippi. On these assumptions relict populations of some species would be found today in places where they escaped destruction by ice, e.g. on nunataks, mountain peaks that remained clear above the ice-sheets. Other refugia could exist in lowland situations between glacier tongues or the lobes of an irregular ice margin, or on any ground by-passed by the glaciation. However, this "no migration" argument would not explain how enclaves of boreal forest came to occupy the higher Appalachian summits beyond the southern limit reached by Pleistocene ice-sheets. The first hypothesis, allowing migration in advance of the spreading ice-sheets — perhaps initiated by climatic cooling before any sign of glaciation — can reasonably explain disjunct distributions such as this.

The periods of glaciation provided an opportunity for cold-tolerant plants and animals to expand their distribution, while exerting a restrictive effect on warmth-requiring (thermophilous) species. It is true that cold-tolerant types were displaced from the areas covered by continental ice-sheets and mountain glaciers but the polar ice did not spread uniformly southward and large areas in east Siberia and Alaska remained unglaciated (Fig. 11.4). On the other hand, warmth-requiring species of the temperate zone were forced to migrate southward into latitudes relatively remote from the influence of polar ice. Here the sub-tropical vegetation and fauna did not give way to the same extent with the result that temperate species had to occupy a reduced area. In Europe the geographical configuration was such that southward migration of thermophilous species led them into several great peninsular blocks separated by seas, namely Iberia, Italy, the Balkans and the Caucasus. Probably as a result of the very different soil conditions already developed there under a warmer climate and because of the new mixture of competing species which they encountered, not all of the southward-migrating species survived the early Pleistocene and others did so only in very restricted areas which qualify as refugia. The

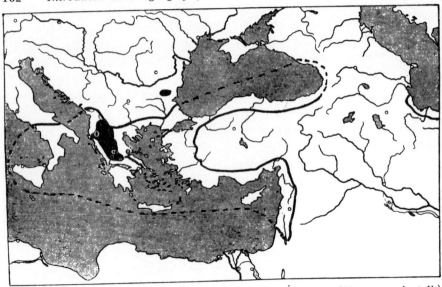

9.9   Arcto-tertiary relicts in southern Europe: (i) Plane-tree (*Platanus orientalis*), distribution outlined; (ii) Horse-chestnut (*Aesculus hippocastanum*), distribution black. (After Rikli, 1943.)

greatest number of species to survive the glacial periods were located in the Balkan peninsula and in the Pontic region (southern and eastern shores of the Black Sea) and, for reasons that are complex and obscure, some of them have remained within the confines of these regions and did not succeed in migrating northward again during the interglacial periods. The Balkans are rich in endemic species and many of them can be accounted for in this way. For example, the Horse-Chestnut (*Aesculus hippocastanum*), a large-leaved deciduous tree with heavy fruit that could only be dispersed by animals, occurs naturally only in two small areas in the Balkans (Fig. 9.9). That it is certainly relict can be seen from the fact that the other trees related to it, i.e. of the same genus, occur only in eastern North America and in eastern Asia from Himalaya to Japan. The absence of closely related forms in neighbouring territory precludes the possibility that it is a recently evolved species. A comparable example, *Rhododendron ponticum*, has been described already and it would seem to have survived the glacial period in Iberia and in the Pontic region, but not in the Balkans. There are two other species of Rhododendron in Europe but both are dwarf shrubs of the high mountains whose entirely different ecological tolerance must have demanded quite the opposite migrational behaviour during the Pleistocene.

The mountain species of *Rhododendron* illustrate the situation of very many plants and animals with montane relict distribution at the present time (Fig. 9.10). In this group we include species whose entire distribution has this form and those which also have a wide area of more or less continuous occurrence in the arctic and sub-arctic. Of the

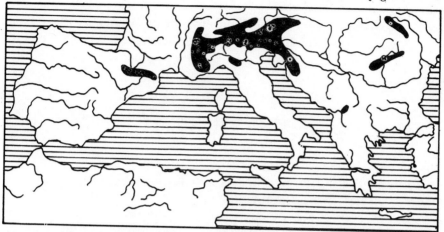

9.10 The disjunct montane distribution of *Rhododendron ferrugineum*, Alpine Rose, in southern Europe. (After Rikli, 1943.)

two alpine Rhododendrons, one is a calcicole species (*R. hirsutum*), the other calcifuge (*R. ferrugineum*). Both have a great altitudinal range (e.g. 600-2600 m and 1200-2850 m respectively) and so their distribution within each mountain group is more continuous than that of most mountain species. It is unlikely that they are confined to mountains by climatic factors: more probably conditions of shade and soil within the needle-leaved forest and above it are suitable but the different conditions of broad-leaved forest are not tolerated. The time when these species could develop wide distribution between the Pyrenees, Alps, Carpathians, etc. would be when broad-leaved forest did not occupy the lowlands, i.e. during the several phases of glaciation. For this reason they are referred to as glacial relicts — the custom is to denote the period at which they had continuous distribution — but it is clear that glaciation did not cause their reliction. It was the increasing warmth of the postglacial period and its attendent changes, e.g. forest migration, that reduced and fragmented their area of occurrence and caused them to retract to higher elevations. Undoubtedly in this process they colonized some ground previously covered by glacier ice. Many species that have arctic as well as montane areas today migrated over ground laid bare by the down-wastage of the northern ice-sheets. It now becomes clear that many of the places where relict populations of mountain and northern species survive today, e.g. on the cliffs of cirques, were in fact buried under glacier ice at the time these species were more widespread. We therefore have the paradox that the places now occupied by glacial relicts are not necessarily the glacial refugia! They may be refugia in the sense that there species have survived the increased warmth and the generally forested landscapes of the postglacial period. Such places include mountain cliffs at high elevations, especially in cirques like Cwm Idwal, north Wales (Godwin, 1955) and also other situations where some combination of circumstances has prevented encroachment by forest. Plants of the glacial period, which are generally intolerant of shade, have survived thus in enclaves far away from those regions where appropriate conditions are widespread.

British refugia and relics

The most remarkable refugia of this type in the British Isles are Teesdale in the northern Pennines and the Burren in County Clare, Ireland. Neither of these are montane situations although the Teesdale refugium is surrounded by high plateau country. The species to be recognised as relict in these places include a bewildering mixture of geographical elements. For the sake of illustration only three plants will be mentioned. Shrubby Cinquefoil (*Potentilla fruticosa*) occurs on river banks and gravels in part of Teesdale. Its other stations in the British Isles are the Burren, a tract of limestone pavement near the west Irish coast, and a montane location in the English Lake District. The differences between the environments of these three places make it difficult to see what ecological conditions they have in common except the absence of forest cover and the possession of base-rich soils in which mineral nutrients are constantly replenished by weathering. Geographically they present a highly dispersed British and Irish distribution of the type noted in Chapter 2. In the wider context, this plant has only few limited occurrences in the whole of Europe (Fig. 11.2) and appears to be relict even in the Ural Mountains. Its continuous area lies in central and eastern Siberia between the Ob River and the Baikal region but in North America it is transcontinental in subarctic latitudes. It is absent from Scandinavia except for lowland sites in southern Sweden.

A geographic parallel among the plants of the Teesdale refugium is the Spring Gentian (*Gentiana verna*), also present in the Burren and with a relict distribution in the mountains of central and southern Europe – but not in Scandinavia which, of course, was the centre of the great ice-sheets that spread south almost to the Carpathians. Thus in the Fennoscandian region great distances had to be covered in northward migration at the end of the glacial period by species intolerant of forest shade. In contrast the ice-margin in Britain was no more than 100 miles south of these refugia where unforested or thinly forested conditions have prevailed throughout the postglacial period.

Plants with a clearly montane distribution in Europe today, such as Mountain Avens (*Dryas octopetala*), must have colonized their present locations during or soon after the glacial retreat and presumably were able to migrate more rapidly. This particular species has a coherent montane distribution in Britain, also occupying favourable sites in

Teesdale and being an important member of dwarf shrub heaths on shell-sand down to sea level in the Hebrides. This is not the highly disjunct distribution to which the last two species were referred. Within Europe as a whole it has its largest area in Fennoscandia where it reaches considerable elevations as well as penetrating far into the arctic latitudes.

South of the limit of glaciation in Britain there are species which seem to have made little or no migrational progress since that time and which have been restricted by the postglacial advance of forest to those few stations where they are now relict. There are few places in southern England where forest could not grow, among them precipitous cliffs in the limestone gorges at Cheddar and the cliffs in the lower courses of the River Avon near Bristol and the River Wye below Hereford. To these can be added certain wind-exposed coastal headlands such as Berry Head (Devon), Brean Down (Somerset) and the Lizard (Cornwall) where there are also skeletal basic soils. These, then, are the postglacial refugia of the lowland zone.

Besides the completely disjunct distribution types that defy alternative explanation, very many species are relict in at least some part of their range. This concept is important because it emphasizes the historical continuity of plant and animal distribution and reminds us that historical influences may be responsible for present boundaries.

SOURCES OF REFERENCE

Bowers, Nathan A. (1965) *Cone-bearing Trees of the Pacific coast.* Pacific Books, Palo Alto, California.

Cain, S.A. (1944) *Foundations of Plant Geography.* Harper, New York.

Elkington, T.T. (1963) *Gentiana verna* L. Biological flora of the British Isles. *J. Ecol.* 51, 755-767.

Elkington, T.T. & Woodell, S.R.J. (1963) *Potentilla fruticosa* L. *J. Ecol.* 51, 769-781.

Elton, C.S. (1958) *The Ecology of Invasions by Animals and Plants.* Methuen, London.

Godwin, H. (1955) Vegetational History at Cwm Idwal: A Welsh plant refuge. *Svensk. Bot. Tidskr.* 49, 35-43.

Hedberg, O. (1969) Evolution and speciation in a tropical high mountain flora, in *Speciation in Tropical Environments,* edited by R.H. Lowe-McConnell. Academic Press, London and New York.

Hubbs, C.L. (editor) *Zoogeography.* A symposium in North American Zoogeography. Amer. Assoc. for Adv. Sci. (1958), Washington D.C.

Hultén, E. (1937) *Outline of the History of Arctic and Boreal Biota During the Quaternary Period.* Stockholm.

Jane, F.W. (1954) The Dawn Redwood. *New Biology.* No. 16, p. 93-100. Penguin, London.

Owen, D.F. (1966) *Animal Ecology in Tropical Africa.* Oliver and Boyd, Edinburgh.

Preston, R.J. (1950) *North American Trees.* Iowa State College Press, Ames, Iowa.

Rikli, M. (1943) *Das Pflanzenkleid der Mittelmeerländer.* Hans Huber, Berne.

Salt, G. (1954) A contribution to the ecology of Upper Kilimanjaro. *Journal of Ecology* 42, 375-423.

Sjörs, H. (1963) Amphi-atlantic zonation, nemoral to arctic. in *North Atlantic Biota,* ed. A. & D. Löve. Pergamon, London and New York

Watson, D.M.S. (1954) Coelacanths, *New Biology.* No. 16, p. 86-92 Penguin, London.

CHAPTER 10
Changes in distribution with time

In the preceding chapters we have found it necessary to refer to historical factors on a number of occasions and in doing so we have implied that environmental conditions have not always been as they are today. If biogeography is viewed only in the present, as the disposition of vegetation or of other ecological communities over the earth's surface in the contemporary condition of the earth's atmospheric circulation, then a quite imperfect idea of the forces that govern the distribution of living things will be formed. Processes take time to exert their effects on the biosphere. The processes themselves are slow in development, measured in terms of a human life-time, and their consequences are felt by plant and animal populations as a whole over several or many generations rather than by the individual. Depending on the average rate of reproduction and on the longevity of each generation, species and communities may possess considerable inertia, which has the effect of delaying their reaction to changes in the environment. For example, we can think of a forest composed of species of long-lived trees. If a change in climate occurs over a period of two or three decades so that the species involved are no longer able to produce and ripen seed, then the regeneration of the forest is imperilled. However, a reversal of the climatic trend some decades later could restore the reproductive capacity of the trees and ensure the perpetuation of the forest. In that hypothetical situation *no change* in vegetation would accompany the fluctuation in temperature, length of growing season, or whatever. Even if the climatic deterioration is not reversed it may have no damaging effects on the standing trees, which continue to function vegetatively though failing to reproduce. Where long-lived species are concerned it may take two or three hundred years before the death of the trees brings to an end the forest vegetation and allows it to be replaced by vegetation of another kind. Note also that all the inhabitants of the forest — the smaller dependent species — are to some extent protected from the direct influence of climatic change because they exist in the microclimate conditioned by the forest itself.

The climate of today is different from that of fifty years ago as meteorological records clearly show. It is easy to dismiss this difference on the grounds that it is of such small amplitude that its consequences will be insignificant. However, the measurable changes in the position of glacier snouts in the Alps and in Alaska over these decades register the differences in a rather impressive way. If we compare the climate of today in the north temperate zone with that of two centuries ago we can see much greater differences. In the period 1684-1814 the River Thames at London froze over in some winters hard enough to allow great fairs to be held on the ice. During the same period the Rhône Glacier extended almost to Gletsch, where it had stood in 1602. All European glaciers advanced to a maximum between 1735 and 1755, a period known as the "Little Ice Age". It is surely true that biological species at the climatic limits of their tolerance will have felt these changes in the peripheral parts of their areas. Since climate (in these latitudes certainly) is not a constant set of conditions, it is reasonable to ask whether some species and some vegetation boundaries really are in equilibrium with the present climate. Their behaviour in response to climatic trends (and other consequential changes, e.g. in soils) over long periods of history, evidently plays a very real part in determining their geographical distribution today.

These are good reasons for insisting that biogeography must take account of where the species we know have lived in past times. Present distributions should be seen as derived from the areas of pre-existing populations which are now preserved as fossils.

The relevant fossils in this context are those that can be identified with species living today or belong to closely related species (denoted by the same generic name) showing evidence of the same ancestry and comparable ecological behaviour. The hazard encountered in dealing with fossils is that the entire animal or plant is only rarely preserved (e.g. Siberian mammoth carcasses in permafrost). Usually only fragments are available and identification must be based on these. Typically, bones, shells and chitinous exo-skeletons are the only remains of animals that are preserved, while isolated seeds, fruits, leaves or wood provide the evidence for plants.

In the plant world recognisable similarity between fossils and living species extends back to the early part of the Tertiary era, approximately 60 million years, but among mammals the rate of evolutionary innovation has been much more rapid with the result that early Tertiary ancestors seem remote by comparison (cf. Chapter 11). Close affinity with modern mammals does not emerge until perhaps mid-way through the Quaternary era, about one million years ago. The time span over which we can trace changes in the distribution of species living now and their recently extinct relatives therefore differs between the flowering plants and the warm-blooded animals, i.e. mammals and birds. Reptiles, fishes and invertebrate groups, however, have a more extended geological scale for their species, like the plants.

Plant distribution in Tertiary times

Fossil floras of Eocene age indicate that plants which we now regard as tropical then grew in the middle latitudes of the northern hemisphere, e.g. in the London basin (51°N) and in the state of Washington (45°N). Representative of this widespread tropical vegetation are the fossils of two palms, Nipa and Sabal, the former recovered from Eocene deposits in the Mississippi basin, the London basin, central Europe, the Black Sea and Egypt. Nipa now occurs only in the Indo-Malaysian region and Sabal only in Florida and the Carribean. Cinnamon trees also had a wide range in the western hemisphere, reaching the latitude of Lake Superior and the south Baltic coast and were found in Patagonia, Tasmania and New Zealand. Today Cinnamons are regarded as characteristically tropical and are restricted to south-east Asia and Melanesia. It was also during the Eocene period that water-lilies of the genus *Nelumbo* had an extended distribution, reaching latitude 55°N in western North America (but not Greenland as reported by some authors) and spread across Europe south of about the same latitude. Now these lilies are separated into an Old World species in Asia south and east of the Caspian and a New World species from the Mississippi to the Orinoco.

Thus fossil distributions, even as old as these, are informative in at least two ways: they show that some plants now confined to certain regions have at some time been circumglobal and they strongly suggest that the climatic conditions of the Eocene were are least sub-tropical up to about latitude 50°N.

In more northerly latitudes during the same period forest of a different kind extended far into the arctic circle and for this reason it has been named the Arcto-Tertiary forest. It does not mean, however, that the climate was then what we call arctic, since the kind of trees that comprised the forest are characteristic of temperate climates today. The plant fossils of this forest belong almost entirely to living genera, i.e. they are specifically distinct from but closely related to species now living.

What is surprising about these widely distributed fossil floras is that they prove the very extensive range of many groups of plants that now have relatively limited geographical associations. In Eocene deposits in Grinnel Land (81° 45′N) in the Canadian Arctic, fossil leaves of Swamp-cypress (*Taxodium*) have been found. The only living species of these trees are confined to the Gulf Coast and southern Atlantic states of U.S.A. and the highlands of Mexico. Associated with *Taxodium* in numerous localities in Alaska, Siberia, west Greenland and Spitzbergen (78°N) were the Redwoods (*Sequoia*), Dawn Redwood (*Metasequoia*) and Maidenhair Tree (*Ginkgo*), all of which we have recognised as geographical relicts by the peculiarities of their modern distributions.

The fossil record confirms this conclusion and shows how far back in the history of the northern continents they attained their originally widespread distribution. Although the trees just mentioned are coniferous there is no analogy with present-day boreal forests, as in the Tertiary era they were intimately mixed with a rich and varied assemblage of broad-leaved trees including sycamores, chestnuts, walnuts, beeches, alders, and oaks. In fact, all the kinds of trees that now characterize the temperate forest regions of the northern hemisphere were already represented in the early Tertiary and, what is more, they had continuous distributions of far greater extent then than most of them possess today. The distribution of fossil and living species of Chestnut (Fig. 10.1) illustrates this and indicates that the existence of separate species in eastern United States, Europe and Japan derives from the circumboreal distribution of that time. In some cases the living species survive in only one or two of these forest regions, though fossil remains prove that they had the same wide range in the Arcto-Tertiary Forest, e.g. Sassafras in the U.S.A.

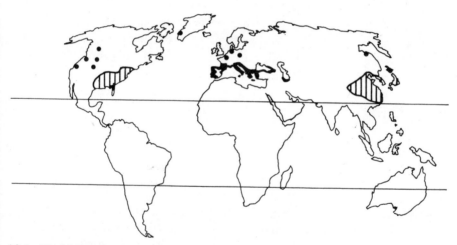

10.1 World distribution of Chestnut (*Castanea*): *C. sativa* in Europe (black), other species in eastern N. America and east Asia. (After Rikli 1943, Meusel *et al* 1965) Tertiary fossil localities: ● (After Berry 1924.)

As the Tertiary era progressed there was a general southward shift of both the tropical and Arcto-Tertiary forests. In the upper Miocene period the forest had lost ground in the arctic but Beech now grew in Oregon and in western Europe in place of the Nipa-Cinnamon flora while Sweet Gum (*Liquidambar*) and Redwood reached the borders of Mexico. Although at this time forests resembling those in temperate regions now moved into the middle latitudes, the great difference remains that forest vegetation was continuous in a broad belt across the American continent. The extinction of Arcto-Tertiary trees in the west accompanied the orogenic activity which uplifted a large area of the continent to higher elevations where coolness of climate made conditions unsuitable. This orogeny continued into the Pliocene period and was accentuated by the uplift of the Cascade Range and Sierra Nevada. This great barrier to the circulation of air-masses from the Pacific made the continental interior arid with the result that forest retreated still further eastward. In Europe the uplift of the Alps during the Miocene did not obstruct the atmospheric flow of westerlies but the east-west orientation of the European mountain chains was to exert a strong influence during the plant and animal migrations of the Pleistocene.

This outline of Tertiary history is very condensed, but it sets the scene adequately for the events of the Pleistocene period. These are known in fuller detail from the abundance of readily accessible deposits of this most recent interval of geological history. From the standpoint of modern biogeography the Pleistocene is especially relevant, because in this period the changing fortunes of the species that populate the earth in its present state can be followed from the distribution of their sub-fossil remains. On a time-scale measured in millenia, the areas occupied by plants and animals are never static and the distribution of fossil remains frequently differs from that of the living populations. Broadly two patterns of change are conspicuous during the Pleistocene: in one group of species the area of fossil occurrences lies to the north of the present range, while in another group the fossil evidences prove that they formerly extended further south than now.

Pleistocene fossils and fossil deposits

Before discussing particular examples of both types, let us look at the character of the fossil materials and the situations in which they have been found. Plant and animal remains dating from various stages of the Pleistocene usually are not mineralized as are older fossils but retain instead their main structural parts chemically unaltered in deposits where conditions for preservation have been good. Thus shell is calcium carbonate, bone remains as calcium phosphate, chitin is unchanged and plant tissues retain their organic hydrocarbons such as cellulose and lignin (the material of wood fibre). Only slowly do they become impregnated with minerals like pyrites and calcite if the beds containing them are subject to percolation by ground-water. The plant fossils of the Cromer Forest Bed in East Anglia were reported by Reid to be pyritized, but these are Middle Pleistocene in age and generally anything younger than this is found to be unchanged. Because of their unaltered condition, in which the actual fabric of the tissues survives instead of being reduced to casts and impressions, these remains are often said to be "sub-fossil".

Bone and shell are preserved best in dry environments — hence loess deposits and caves are often rich in these materials — and in aquatic sediments only non-acid conditions can preserve them. Fluviatile deposits such as flood-plain alluvium incorporates lenses of drifted debris that has settled in backwaters and oxbows. These situations tend to preserve the most varied assortment of materials, both 'plant and animal, since anything that falls or is swept into flowing water may ultimately be deposited in this way. Of course some remains may have travelled a considerable distance downstream before settling and there is a risk that some sub-fossil material may be moved from older deposits upstream by bank erosion and then re-deposited. It is lacustrine sites that are the most favourable for the preservation of organic remains, especially small lakes, ponds and pools which receive water-transported debris only from the immediate surroundings. Quiet water conditions and continuous sedimentation lead to the accumulation of underwater muds where oxidation has little time to act. Species that inhabit such sites will naturally be among the most frequent sub-fossils in aquatic muds, but leaves, fruits, seeds, branches of trees, insects and snails may also be carried or blown into the water and eventually incorporated in the sediment.

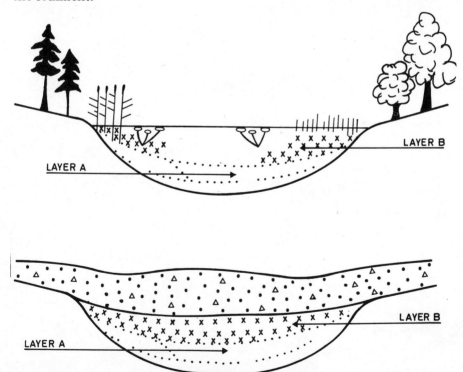

10.2 Infilling of lake-basins by accumulation of plant detritus and other sediment Layer A: fine sediment formed while open water conditions exist. Layer B: coarse detritus deposited in reedswamp. Lower diagram shows a completely filled basin sealed by later non-lacustrine deposits.

Within the context of the Pleistocene period some types of site are relatively permanent features of the landscape (caves and deep lakes) while others are relatively transitory (e.g. river back-waters and small ponds). Examples of the latter group originate in every temperate phase of the climatic cycle (i.e. in every interglacial period) and long after the pond has filled up the lacustrine deposits may be buried beneath materials from extraneous sources under the influence of changed climatic conditions (Fig. 10.2). Thus they are sealed in by loess or soliflucted sludge or even by glacial till — for in a frozen state old lake deposits have been overrun by glacier ice without serious disruption. All these circumstances need to be known in order to assign the fossil contents of sedimentary deposits to the appropriate phase of the Pleistocene and avoid confusion between the distribution of species during distinct episodes.

We turn now to the actual distribution areas of species whose sub-fossil remains are frequent in Pleistocene deposits. If the lists of species identified as sub-fossils from many interglacial sites are assembled, maps can be prepared to show the geographical distribution

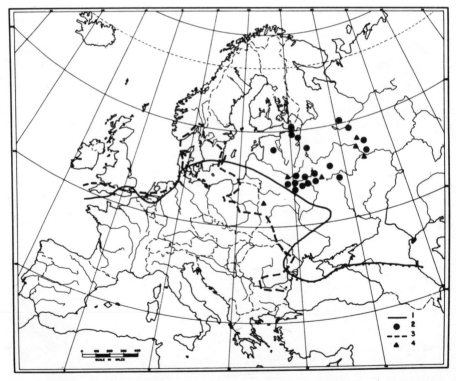

10.3   Interglacial shifts of European trees. (1) Present northern limit of Hornbeam (*Carpinus betulus*). (2) Interglacial fossils of Hornbeam. (3) Present northern limit of Beech (*Fagus sylvatica, F.orientalis*). (4) Interglacial fossils of Beech. (From Butzer 1965, modified).

of records for the more frequently represented species. Because the location of interglacial deposits, e.g. old lake beds, is usually not detectable at ground surface (cf. Fig. 10.2), the discovery of fossilferous organic layers depends on chance exposures made in the course of quarrying for gravel and brick-clay or in trenching for pipelines or road cuttings. For this reason the distribution data are skeletal and are therefore best represented by dot maps. Since more material of interglacial age is available in Europe than in North America, the examples that follow are European.

Northward Pleistocene migrations

In sites where there has been continuous deposition of sediments there is evidence that the intervals between successive glacial episodes were of long duration. There has therefore been time enough for the processes of migrational advance to re-instate species in areas that were intermittently glaciated, a fact to which the distribution of sub-fossils bears witness.

In interglacial times European Beech and Hornbeam, in common

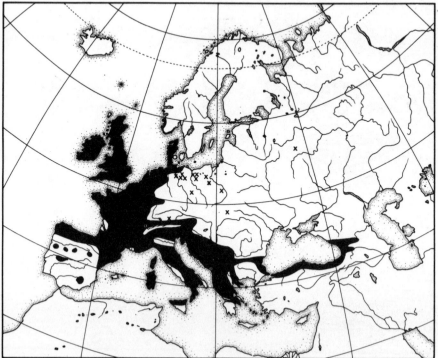

10.4   Distribution of European Holly (*Ilex aquifolium*) and of interglacial fossils of the same species: X (After Frenzel 1968.)

with many other broad-leaved forest trees, achieved a considerably more extensive distribution than they have at the present time (Fig. 10.3). Evidence from the Scandinavian countries is lacking because subsequent glaciation has removed the deposits formed in interglacial lakes, but in Russia sub-fossil remains of *Fagus* and *Carpinus* prove that their distribution reached latitude 60°N and longitude 45°E. Both species which now occur only west of the Crimea penetrated much further into the heart of the Eurasian continent under interglacial conditions. Thus an eastward trend in tree migration accompanied their northward expansion. This is also seen from the sub-fossil distribution of Holly (Fig. 10.4) but because of the erosion of interglacial deposits to the north of the Baltic there is no indication how far north it grew.

In recognising the fact that many plants were more widespread during the temperate phases of the Pleistocene we are immediately faced with a problem of interpretation. Is it because they had a longer period of time in which to spread during the interglacials than has elapsed since the end of the last glaciation? If so, we should expect to

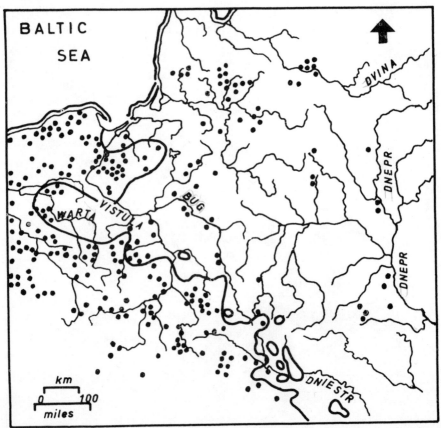

10.5   Present eastern limit of Beech (*Fagus sylvatica*) in Europe (continuous line). Places named after Beech shown as dots (by Turowska, 1929 and Szafer, 1952). A former extension to the Dvina and Dnepr Rivers is indicated.

find that these species are still migrating and that their present boundaries do not represent the climatic limits of their tolerance. Against this view there is evidence that Beech had in recent times advanced further east in Poland than its present limit and has since retreated somewhat (Fig. 10.5). This seems to indicate that the tree had reached an effective climatic limit and has therefore no potential for further migration under present conditions. The existence of climatic control at the northern and eastern limits of Holly has already been discussed (p.79) and the observed damage to this tree caused by the exceptionally severe winters of 1939-42 dispels any doubt on this question. On the other hand is it right to assume that the limit of climatic tolerance of Beech or Holly is the same now as it was many thousands of years ago or at least that it has not changed significantly? This assumption underlies all climatic reconstructions built upon the paleo-geographic evidence of fossil distributions. The argument in support of its validity is that if the ecological tolerances of species have changed during the time-span of the Pleistocene then all temperate species have changed their behaviour in the direction of reduced tolerances, which seems improbable. It is difficult to escape the conclusion that the interglacial periods experienced adequate rainfall for temperate forest over wider areas and were warmer than the present climate of Europe.

This conclusion is confirmed by the distribution of sub-fossil remains of animals independent of forest environments. The freshwater mollusc, *Corbicula fluminalis,* has regularly been found in interglacial deposits of

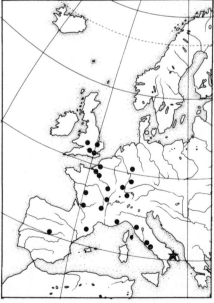

10.6 Distribution of *Hippopotamus* in western Europe during the Pleistocene. Fossils assigned to interglacial periods: ●; and to the Weichselian glacial period ★. (After Frenzel 1968.)

rivers and estuaries over a large area of Europe, reaching as far north as Thuringia, Halle and Flaming in central Germany, as well as the Netherlands, Belgium and northern France (Charlesworth, 1957). In England it has been found in interglacial deposits of the Thames between Ilford and Tilbury and in East Anglia at Clacton, Harwich and Cambridge (Zeuner). In Europe today it is only found in Transcaucasia (cf. *Rhododendron ponticum*) and its next nearest locations are in western Asia, e.g. the Jordan Valley and Turkistan (U.S.S.R.), and in north Africa. If this distribution can be regarded as climatically controlled then greater summer warmth in northern Europe is indicated for the interglacial periods, the only qualification with regard to winter temperatures being that the January mean would be above freezing-point in the parts of Europe where *Corbicula* occurred. These climatic implications are supported by the interglacial distribution of *Hippopotamus* in Europe, as mapped by Frenzel, which is closely comparable to that just described. The bones of this mammal prove that it lived as far north as East Anglia in the more oceanic west of Europe but records east of the Rhine seem to be lacking (Fig. 10.6). The general pattern of recurrent northward extension of distribution during interglacials is repeated in many species of water plants such as Water Soldier (*Stratiotes*), Frog-bit (*Hydrocharis*), the water fern *Salvinia natans* and the pondweed *Naias minor*.

The discussion so far has included species whose present distribution places them into one or other of the broad endemic types* e.g. the nemoral and the sub-atlantic elements within the continent of Europe. Evidently these plants and animals spread and occupied approximately similar areas in each of the several interglacial periods (thus verifying their temperate climatic character) and they occupy only somewhat smaller areas of the same general form at present. However, there are other species whose changes in distribution have not been so regular or repetitive. On the basis of present range they can be characterized as regional endemics rather than as broad endemics. Their total areas are small and are generally concentrated within — if not actually confined to — individual physiographic regions, e.g. the Iberian, Balkan, and Karpato-Helvetic regions (the term Alpine is open to objection because it is commonly used to designate an altitudinal zone above tree-line and strictly it does not include other central European mountain groups). Additionally, species whose Pleistocene history has been irregular include examples of relics in which disjunct and extremely reduced areas now stand in marked contrast to the distribution of their sub-fossil remains from earlier temperate periods. In both of these types

---

* *Endemic distributions* are those in which the species, family or whatever group is contained entirely within one natural region, country or continent. We can distinguish broad endemic ranges, covering an extensive territory, and narrow endemic ranges which are more restricted e.g. a peninsula, an island or mountain range.

species attained wide distribution in only one or perhaps two interglacials, not necessarily the same periods in all cases, and in other interglacials they failed to achieve the northward migration. It follows that regional endemic and relict species contribute some of the most distinctive elements to interglacial floras and provide a means of identifying deposits formed in separate temperate periods. In ecological terms they were important in modifying the appearance, form or functional aspect of the vegetation, as for instance when they included a larger proportion of evergreen species or of ericaceous members, and they possibly serve to differentiate the climatic conditions of one interglacial from another.

Plants whose distributions are now largely centred on particular physiographic regions but which spread widely outside those areas in particular interglacial periods include the maple *Acer monspessulanum*, Box-Tree (*Buxus sempervirens*), Silver Fir (*Abies alba*) and the heath *Erica scoparia*. The modern distribution and Pleistocene history of these species illustrate a relationship between the two.

During the last interglacial period *Acer monspessulanum* was a frequent component of mixed deciduous forest across Europe, reaching at least as far north as East Anglia (Ipswich) and the London Basin (deposits on the site of Trafalgar Square!) Its winged fruits, similar in general appearance to those of other maples, are abundant among the macrofossils at some sites of this temperate stage but apparently not in earlier interglacials. At present it can be characterized as a Balkan element (Fig. 10.7) with outlying occurrences in peninsular Italy, Sardinia, and Sicily, but it is altitudinally divorced from the Mediterranean climatic domain. In the northern Adriatic area (e.g. Istria) it is a typical member of the Manna Ash-Deciduous Oak Woods where associated trees are a mixture of nemoral species (*Quercus petraea, Carpinus betulus, Acer campestre*) and predominant south European species (*Quercus lanuginosa, Q. cerris, Carpinus orientalis, Corylus colura, Ostrya, Fraxinus ornus, Celtis australis*). Woods of this type occur in the lowland and hill zones of the western Balkan peninsula up to 800 m. Further south *Acer monspessulanum* is restricted to the higher elevations (600-800 m) as a constituent of the Oak-Chestnut forests of the lower montane forest zone, i.e. below the Beech-Fir forests but above the sclerophyllous evergreen woodland and scrub of the Mediterranean littoral. There are a number of small disjunct areas in its modern distribution (Vendée, Côte d'or, Jura, Savoy, Middle Rhine and south Tirol) that are certainly relict but these probably reflect postglacial changes in its territory and are not survivals from the interglacial.

In the preceding interglacial (the penultimate) Box-tree and Silver Fir migrated with other trees to form forests of mixed or even predominantly evergreen character in regions far to the north and west

Acer monspessulanum

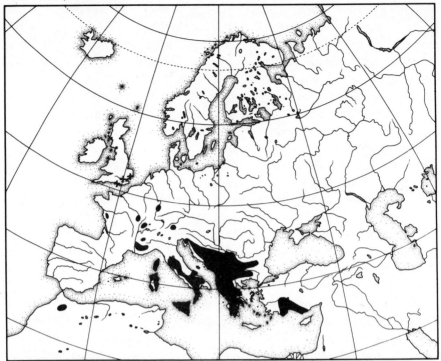

10.7   Modern distribution of the Balkan Maple (*Acer monspessulanum*) from data of Chadefaud & Emberger, Davis, Hegi, Rikli and Turrill. Also, northernmost fossil localities of last interglacial age (Ipswichian) from West 1957, 1960, 1964.

of their present-day distribution. Pollen of Box-tree and needle-leaves of Silver Fir have been found in west-central Ireland at Gort together with sub-fossils of other evergreen species e.g. Holly and Yew. Silver Fir (*Abies alba*) is today a central European montane species (Fig. 10.8), usually occurring in the middle altitudinal zones together with Beech. Its distribution is considered by Firbas to be climatically conditioned and its absence from the sub-atlantic domain indicates that it is sensitive to spring frosts, as experience has shown in plantings for afforestation. The elevations at which it occurs are dictated by its need for plentiful rainfall rather than by temperature limits and for this reason it ascends in north Macedonia to between 1500-1700 m where rainfall is well distributed throughout the year. Despite the southern latitude here the mountains in this zone are snow-covered from mid-November until· mid-March. Box-tree (*Buxus sempervirens*) has a disjunct Cantabrian-Balkan-Pontic distribution which suggests relict status although the individual territories within each of these areas are considerable (Fig. 10.9). It is especially characteristic of the pseudomacchie of Albania and north and south Macedonia. This vegetation is xerophilous evergreen brushwood or scrub which inhabits the submontane and montane zones but not the lowland Mediterranean

Abies alba

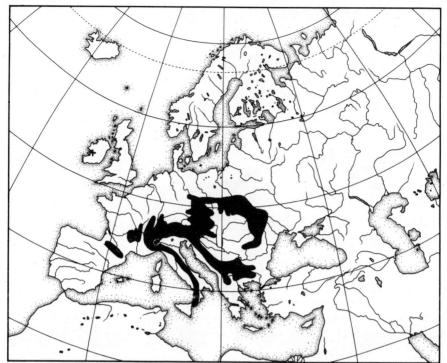

10.8   Modern distribution of Silver Fir (*Abies alba*). (After Mattfeld in Rubner 1934) Also, fossil locality of penultimate interglacial age at Gort, Ireland (Jessen *et al*, 1959).

districts. Incidentally Holly and Ivy are among the associated species in this scrub and together with Box-tree are frequent in the understory of montane forests (deciduous oaks, etc.) in the Balkans. The wide temperature tolerance of Box-tree is illustrated by its presence at elevations up to 2000 m on the Thessalian Olympus. Thus a temperate regime with warm summers and abundant rainfall seems to characterize all these formerly extended interglacial species.

Another quite distinct geographical element in the flora of the penultimate interglacial in Ireland is represented by Iberian species. Plants belonging to this group have no overlap in their modern distribution with central European types. *Erica scoparia* is a heath plant of south-western Europe whose sub-fossil remains show that during this interglacial it extended northwards in oceanic districts to Ireland and the Shetlands (Fig. 10.10). A comparable example showing the advance of south-western species in an earlier interglacial (the Cromerian) is the poppy-like plant, *Hypecoum procumbens*, which was found fossil in East Anglia but now is described as "atlantic-mediterranean" (Rikli). It is a spring-time ephemeral herb of sandy places along and near the littoral.

Among species which now have totally relict distributions there are some that occupied more extensive, probably continuous areas during

10.9 Modern distribution of Box-tree (*Buxus sempervirens*) in Europe (shown black with scattered localities as dots) after Christ 1913, somewhat simplified. Also, fossil locality of penultimate interglacial age ▽ in Ireland (Jessen *et al*, 1959).

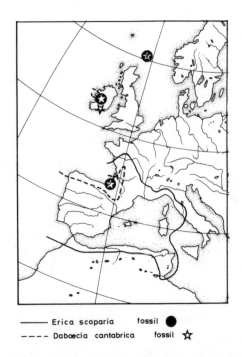

——— Erica scoparia   fossil ●
– – – Daboecia cantabrica   fossil ☆

10.10 Iberian and Atlantic elements of the Gortian interglacial (penultimate). Modern distributions after Hanson, 1950 and Good, 1947. Fossil records from Birks, Jessen, Oldfield and Watts.

one or another of the Pleistocene interglacials. *Rhododendron ponticum* probably had a wide European range in the penultimate interglacial when it was associated with several other evergreen shrubs and trees in luxuriant forests of mixed composition. Proof of its occurrence in Ireland, south-west France and Austria rests upon identification of sub-fossil capsule valves (of the fruit), leaves and pollen grains in deposits at Gort, Biarritz and Hötting in the Inn valley. This fossil record does much to explain its modern distribution in the Pontic and Iberian regions. How much may be safely inferred from the latter in reconstructing the climatic conditions of the inter-glacial is more doubtful. It would certainly be wrong to characterize the Pontic region as "continental" for the proximity of the Black Sea ensures ample precipitation, high humidity and moderation of winter temper-atures. As we know that this plant now flourishes in Ireland as a naturalized escape from cultivation, we cannot regard the summer warmth of the Pontic area as essential for its success.

A eu-atlantic group of relict species also occurred over an extended territory during the penultimate interglacial, certainly spreading northwards along the Atlantic seaboard to reach the Shetland Islands and probably occupying the whole of Ireland. Examples here are two ericaceous plants, St. Dabeoc's Heath (*Daboecia cantabrica*) and Mackay's Heath (*Erica mackaiana*).The first has a narrow endemic range in the Cantabrian region of northern Spain while the second is virtually restricted to the province of Asturias within that region (Fig. 10.10). Both have small disjunct areas in the extreme west of Ireland that we recognise as relict and presumably both species must have survived near the western margins of the continental shelf since interglacial times. The Irish localities where there are existing populations of these heaths possess no unique climatic character to set them apart from other western peninsulas where neither species is present, and there is no apparent reason why *Erica mackaiana* does not grow to the west of Asturias in the Atlantic peninsulas of Galicia. Yet another species, Cornish Heath (*Erica ciliaris*), which is now absent from Ireland, is proved to have grown there during the penultimate interglacial by sub-fossils found at Gort, County Galway and at Kilbeg, County Waterford. This particular interglacial period gains its distinctive character from the general presence of trees with evergreen foliage and, in westernmost Europe, from heath communities representing the increased importance of eu-atlantic species. These are now much reduced for reasons involving their rates of migration, the time necessary for appropriate soil types to develop and the rate at which sea-level rose with the release of water from the polar ice-caps. There are no signs that any of these plants succeeded in recovering their extensive areas during the last interglacial period, nor have they done so in the course of the postglacial period. It would appear that the exigencies of the two glacial periods that have passed since their wide distributions were attained drove the surviving populations southward and westward and the opportunities for their dissemination and establishment were never repeated. For all of the species discussed so far in relation to the northward extension of range the sub-fossil occurrences are situated in the deposits of ponds and shallow lakes that existed during one or other of the interglacial periods, i.e. in the stratigraphic position represented in Fig. 10.2b. Those lakes no longer exist and the muds that filled them are overlaid by deposits of aeolian, fluvial or glacial origin related to the cold period which followed the interglacial.

Postglacial northward migration

Evidence for northward migration beyond present distribution limits can be found for many species of plants and animals whose sub-fossil remains are postglacial in age, i.e. situated in the muds of lakes that are features of the present landscape as Fig. 10.2a. From the lower or middle layers of such lake sediments the identifiable remains are

Corylus avellana

10.11 The distribution of Hazel (*Corylus avellana*) in Fennoscandia during the mid-postglacial period and at present. Location of sub-fossil nuts at sites north of present limit: X. General limit of present distribution shown as continuous line, additional scattered localities as dots. (after Hultén 1950)

sometimes found of species which at present live nowhere near the site and whose nearest living populations are either further south or at lower altitude. Again the most frequent examples are either forest trees or water plants. The unmistakable nuts of Hazel and the highly distinctive spiked fruits of Water Chestnut both occur frequently in bottom mud of lakes in Finland and Sweden up to 200 miles (320 km) north of their present limits. Hazel does occur in the southern parts of both countries but previously extended to about latitude 64°N (Fig. 10.11). Water Chestnut is not at present found in Finland but frequent records of its sub-fossil nuts prove that it was widespread in southern Finland at an earlier stage of the postglacial (Fig. 10.12). In south Sweden it is still present at three sites but their dispersed distribution and the much greater number and spread of fossil sites indicate that the plant is a hypsothermal relict in Sweden, i.e. the remaining populations are survivals from the phase of greatest warmth within the postglacial period, when its more extensive distribution was achieved. The pond-weed, *Naias flexilis*, has similarly changed its distribution almost in parallel and remains at a few scattered places in south Sweden. The Opposite-leaved Pondweed (*Potamogeton densus*) provides a further example of displacement of the northern limit during the postglacial. This pondweed has its limit of continuous distribution

Trapa natans

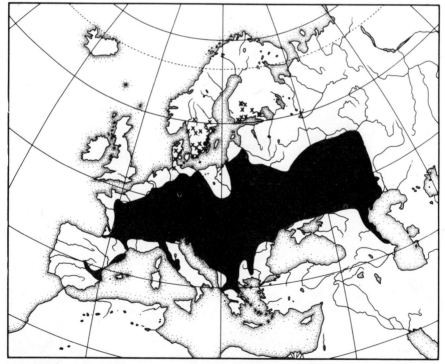

10.12 Modern distribution of Water-chestnut (*Trapa natans*) (after Gams, in Walter 1954) and location of sub-fossil nuts X from the postglacial period at sites north of its present limit (after Hultén 1950).

in Jutland and in northern England, with no living populations in Sweden. However its sub-fossil fruitstones prove that it had migrated further north into Sweden and that its present boundary represents a retrogression.

There is no doubt that these fluctuations have taken place under the influence of changes in summer temperature. The limit of the distribution area of Hazel during the postglacial warm period very nearly coincides with the present July isotherm of 9.5°C, whereas the present limit coincides with the 12°C isotherm. Consequently we can conclude that July temperature has fallen 2.5°C since the hypsothermal period. Climatic change of this order is also suggested by the geographical distribution of breeding populations of the European Pond Tortoise then and now. The breeding range of the Pond Tortoise lies in central Poland and Germany though non-breeding individuals can be found in a zone immediately to the south of the Baltic Sea coasts. However, very tangible proof that the species was breeding in Jutland at an earlier phase of the postglacial comes from the discovery of its sub-fossil eggs at a number of sites in Denmark. On this basis the hypsothermal period is estimated at 2°C warmer than now on average mean monthly figures for July.

The date of northward expansion of thermophilous species can be determined by radiocarbon assay of samples taken from the stratified lake muds at levels containing the macro-fossils or even from the carbonaceous fossils themselves if a sufficient quantity can be collected.

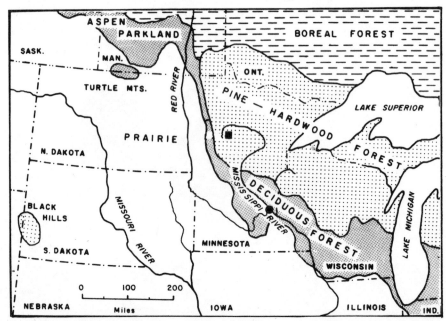

10.13 Map of the forest-prairie border in Minnesota and adjoining states (adapted from McAndrews, 1967), showing location of sub-fossil sites at Martin Pond ▪ and Kirchner Marsh ● (from Watts and Winter, 1966).

Microscopic analysis of the pollen content of stratified sediments is also useful in providing a relative time-scale and in charting the main changes in the vegetation cover around the site of deposition. The layer in which fruits of Water Chestnut (*Trapa natans*) occur in Swedish sites is dated at around 6,200-7,000 B.C. It was formed when forests of thermophilous trees such as Elm and Hazel first began to replace the pioneer Pine in that country.

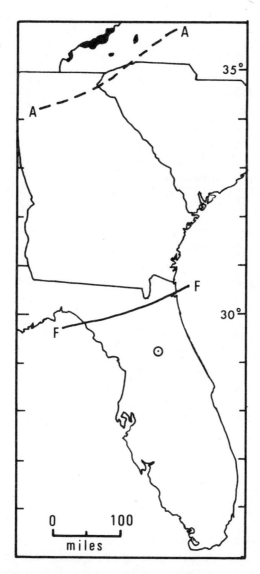

10.14 Plant migration in south-east U.S.A. during the Pleistocene. Modern distribution of Spruce (*Picea*) shown black, and southern limits of Sugar Maple (*Acer*) marked (A) and of Beech (*Fagus*) marked (F). Microfossils of all three occur at Mud Lake, Florida (ringed ○). Adapted from Watts 1969, and Preston 1950.

The advance of the prairie

In North America the shifts in distribution of individual species and vegetation types associated with the hypsothermal phase are best known in the mid-west, where the boundary between forest and prairie represents the point of contact between two distinct floral provinces and a precarious ecological balance exists that is sensitive to climatic change (Fig. 10.13). From the deeper layers of lake sediments in Minnesota and neighbouring South Dakota the pollen record reveals a succession of forest types in the early postglacial period: first spruce forest, then a mixed conifer-hardwood forest including fir and pine with alder, ash and birch, then deciduous forest of elm and oak, which prevailed until about 7,100 years B.P. After this time less tree-pollen in the sediment signifies a retreat of the forest in the vicinity and a larger component of wind-pollinated herbs indicates that its place was taken by prairie vegetation. During this interval of about 2000 years, lasting until 5,100 B.P., the macrofossils then deposited in a number of small lakes that were critically located (Fig. 10.13) indicate what kind of climatic change had come about. The sub-fossil seeds and fruit of more than a dozen annual herbs appear for the first time in the sediments at or soon after 7,100 B.P. and fluctuate erratically in abundance throughout the prairie interval. The plants in question are known today either as weeds (of cultivation) or as ruderals, especially of damp ground, and they are the same species that characterize the exposed mud around lake edges when summer drought results in lowering of the water level (the so-called "draw-down" of the prairie potholes). Their arrival in central and eastern Minnesota was facilitated by the high incidence of dry summers in which lake water level was lowered and temporary habitats suited to these species came into existence seasonally in most years. The end of the prairie interval on the pollen diagram follows shortly after the disappearance of macrofossils of these annual plants at the transition from the middle to upper sediment layers. After 5,000 years B.P. oak forest advanced again from the east and reinstated the area between Martin Pond and Lake Carlson within the forest province after an interlude of grassland dominance.

Southward Pleistocene migrations

Species whose distribution formerly extended south of the territories they now occupy mostly belong to quite different geographical elements from those which migrated northward during the temperate interglacial and postglacial periods. However in certain regions in the low latitudes of the temperate zone, such as the Florida peninsula, there is sub-fossil evidence that even the trees of the temperate forests

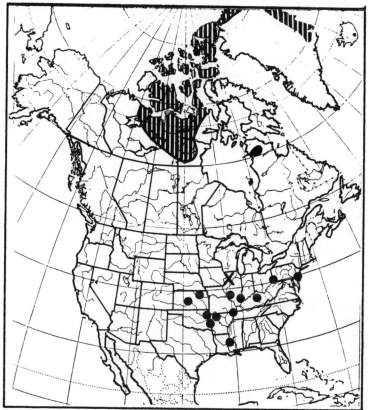

10.15 Present distribution of Muskox (*Ovibos moschatus*) (shaded) and Pleistocene records of this species X, and of the extinct form, *Symbos cavifrons*, ●. (After Blair 1958 with additions.)

migrated southward beyond their present limits at certain times during the Pleistocene (Fig. 10.14). At Mud Lake in Marion County, Florida, lake sediments more than 13 m deep contain a pollen record of forest and other vegetation within the surrounding area over a long period of the Upper Pleistocene (Watts, 1969). The sediments are interrupted by a barren sand layer and above this a radiocarbon date of 8,160 years B.P. at 430 cm indicates that the upper deposits are "postglacial" (though of course Florida was never glaciated). Just below the sand layer another dated sample shows that the lower 8 metres of deposits have an age greater than 35,000 years. Interestingly the lower series contains sub-fossil pollen of Sugar Maple and Beech and the basal 2 metres contains similar evidence for Spruce, none of which are represented in the upper series of sediments. None of these trees grow as far south as this any longer but now reach their southern limits at various distances to the north of Marion County. Nearest is the Beech (*Fagus grandifolia*) with its southern limit in the Florida panhandle, then Sugar Maple (*Acer saccharum*) reaching only northern Georgia and furthest distant is Spruce (*Picea rubra*) now in the Great Smoky Mountains 400 miles away. The fossils of these northern trees in

Florida clearly indicate a period of cooler climate in that area. Knowing the scale of deposition of the sediments and that they precede the "postglacial" period with its quite different warm temperate flora we can surmise that Beech, Maple and Spruce migrated south into the Florida peninsula during the glacial episodes.

The fossils in higher latitudes confirm this interpretation and fill out some details of the picture of glacial times. Sub-fossil wood of Jack Pine, Tamarack, White Spruce and Bog Cedar proves that these boreal forest species extended southward into Louisiana and South Carolina. Finds of Muskox bones in glacial deposits in Illinois demonstrate that this tundra mammal found appropriate conditions in the mid-continent and the more frequent fossils of the extinct Pleistocene form, *Symbos*, show that it ranged over a vast territory marginal to the ice-sheets (Fig. 10.15). Comparable distribution of Reindeer and Mammoth has

10.16 Modern and Pleistocene distribution of *Dryas octopetala* in Europe (after Tralau, 1961 and Perring & Walters, 1962). Present occurrences shown black. Sub-fossil occurrences of the last glaciation (Würm) and its retreat □; sub-fossils older than this ▽ . The southern limits of ice-sheets of the last glaciation (i) and of earlier glaciation (ii) are shown.

been mapped in what is now the temperate zone of both North America and Europe.

Plant species which were widespread where temperate deciduous trees are now predominant include many with present-day arctic and arctic-alpine distributions e.g. Dwarf Birch (*Betula nana*), Least Willow (*Salix herbacea*), Mountain Avens (*Dryas octopetala*), and Alpine Meadow-rue (*Thalictrum alpinum*). Thus both dwarf shrubs and herbs are represented but the fossil assemblages in which they are found usually show no evidence of the contemporary existence of trees. The situations and stratigraphic relations of the deposits containing floras of this type differ from the little disturbed lake-bed deposits characteristic of interglacial periods. Typically the organic beds containing sub-fossils of the northern flora and fauna are broken-off lumps of peat or silty mud deposited in a disorganized way amid great thicknesses of river gravels. They were probably first transported and then embedded while in a frozen condition. The map for Mountain Avens (Fig. 10.16) illustrates the characteristics of an arctic-alpine distribution and the sub-fossil occurrences of this species, and it is typical of many other plants assigned to the same element today. It is generally present at low elevations the arctic and in the hyper-oceanic conditions of western Ireland, west Scotland and Iceland. The plant grows at increasingly high elevations in the Scandinavian mountains as it is followed southwards. It also has widely disjunct populations in many of the mountain ranges of central and southern Europe. In this respect the arctic-alpine distribution can be identified as a geographically relict type and its disjunction can be explained in the light of fossil evidence.

We may recognise exactly corresponding examples among beetles, often well represented in Pleistocene peat and pond deposits because their tough exo-skeleton resists abrasion and decomposition. The distribution of the carabid beetle, *Amara alpina*, (Fig. 10.17) corresponds closely to that of the dwarf shrub *Dryas octopetala* while other beetle species match other plants having somewhat more extensive (boreal) distribution, e.g. *Patrobus septentrionalis*, or less extensive (arctic and arctic-montane) distributions, e.g. *Bembidion tellmanni*, *Simplocaria metallica*, *Diachila arctica*. The sub-fossil distribution of Mountain Avens shows that the plant flourished and was probably an important member of the vegetation south of the last ice-sheet at its maximum extent (cf. localities in south Poland, the alpine foreland and southern England). Sub-fossil remains of all these beetle species have been discovered at sites outside the glacial limit, showing that they too occurred far to the south of their present limits. As the ice-sheets retreated northward during the waning phases of the glaciation both plants and beetles must have migrated into the territory exposed, since many *Dryas* sub-fossil localities are within the area ice-covered at glacial maximum (Fig. 10.16). It becomes apparent that the present

10.17 The northern or arctic element in the glacial fauna and flora of Europe. (a) Modern distribution of the carabid beetle, *Amara alpina*. (b) Modern distribution of the dwarf shrub, *Dryas octopetala*. Crosses mark fossil localities dating from the last glacial period (Weichselian). (a) After Coope 1962; (b) after Tralau 1961.

distribution areas of high-arctic and sub-arctic species in the European sector (i.e. Fennoscandia and east to the Urals) and in the Hudson Bay sector of North America are very recent since they occupy ground where the latest vestiges of ice-caps remained until 10,000 years ago (and even less locally). Even in the mountain areas further south the present sites of "alpine" species have been colonized within the same period of time, for certainly at their maximum mountain glaciers descended below the present timberline and the nival zone of that period blanketed the higher regions where their populations survive today.

The associations of sub-fossils dating from the Pleistocene cold periods also include species belonging to other geographical elements whose presence at this time might not be expected. Prominent among these are species with markedly continental distribution today, i.e. eastern European. In many cases their present northern limits are no more extensive than those of the thermophilous species that character-ized the temperate interglacial periods! The examples of *Euporocarabus hortensis* and *Tomocarabus convexus* (Fig. 10.18) illustrate the dis-tribution of beetles which previously extended westward, as proved by sub-fossil finds at periglacial localities in Britain. Continental elements in the periglacial flora of the north European plain, which was then continuous across the southern North Sea, are exemplified by finds at Colney Heath, Hertfordshire, including the distinctive pollen micro-fossils of *Ephedra* cf. *distachya* and of Sea Buckthorn (*Hippophaë rhamnoides*). Belonging to the so-called Turanian element, their main distribution is now in the region of the Aral and Caspian Seas (Fig. 10.18), the "mid-west" of Eurasia, and ecologically they are plants of steppe and desert. West of this main area both are represented by relict distributions in Europe. *Ephedra* occurs at small highly dispersed localities in dry and warm, sandy or stony habitats, e.g. in valleys of the Alps in Dauphine, Valais and south Tirol up to 750 m and along the French Atlantic coast. *Hippophaë* also occurs in the Alps within the forest zone (i.e. below the tree-line) where it occupies the fresh and disturbed soils of landslips and other unstable situations. This spiny shrub has an entirely separate (disjunct) area around the coasts of the North Sea where it colonizes sand-dunes and cliffs of boulder clay. What is enlightening about these examples is the way in which the widespread sub-fossil distribution makes the anomalies of the present situation intelligible rather than vice versa. Clearly during glacial periods these species found quite suitable environments extensively in Europe south of the ice-sheets and since the last glaciation they have been eliminated from most of that area by the spread of forests, as they do not tolerate shade. They did not migrate northward after the glaciation, as did the arctic species, and climatically it would seem that strong insolation and seasonal warmth are necessary during their period of

10.18 The eastern or continental element in the glacial fauna and flora of Europe. (a) Modern distribution of the carabid beetle, *Tomocarabus convexus*. (b) Modern distribution of the steppe plant, *Ephedra distachya*. Crosses mark fossil localities dating from the last glacial period (Weichselian). (a) After Pearson; (b) after Meusel.

growth. These characteristics of climate are functions of latitude and would be unaffected by the development of ice-sheets in the region.

Another more northern continental element represented widely by its distinctive pollen grains in late glacial pre-forest environments is Jacob's Ladder (*Polemonium coeruleum*). It is a tall herb found in hillside meadows, forest margins and even stony and rather dry habitats within the montane forest zone from the Balkans to Scandinavia. Thus while it is intolerant of forest shade it is confined to altitudes below timberline, e.g. 1200 m in southern Norway (Hultén) and does not reach subalpine elevations even in the Balkan mountains (Turrill). Consequently it now has a fragmented distribution in Europe, being absent from the lowlands, and in Britain it is clearly relict in scattered limestone localities at moderate altitudes. These must be refugia in and around which the plant has persisted since late glacial times. The main part of its distribution is described as west Siberian and it is found in the boreo-nemoral zone, reaching about 60°N in Russia (and further north in maritime Scandinavia). Although tolerant of fairly severe winter cold, it also requires sustained warmth during the growing season, and this distinguishes it from arctic and sub-arctic plants.

Lastly the vegetation of Britain and the neighbouring continent peripheral to the Pleistocene ice-sheets included species that have now all but disappeared from Europe. Their few relict stations appear so diverse in character and so erratic in geographical location that they would be impossible to explain if it were not for the presence of sub-fossils. These disclose that the plants existed in glacial and late glacial times over some of the intervening territory at least. Shrubby Cinquefoil (*Potentilla fruticosa*) is a case in point. Now east Siberian — North American in its continuous distribution, it is absent in Laurentia (Hudson Bay) and Fennoscandia, precisely where the ice of the last glaciation remained longest (cf. Fig. 11.4).

One of the most instructive discoveries to emerge from study of sub-fossil remains and their palaeogeography is that plants no longer associated either ecologically or geographically formed types of vegetation that were widespread in unforested landscapes at various times in the Pleistocene. The frequency of suitable habitats in the changing landscape after the Ice Age, the migratory capacity of species and the opportunities open to each one, as determined by its ecological tolerance, are factors that have played a decisive role in shaping the territories occupied by plants and animals today.

SOURCES OF REFERENCE

Birks, H.J.B. and M.E. Ransom (1969) An interglacial peat at Fulga Ness, Shetland. *New Phytologist* 68, 777-96.
Chadefaud, M. and Emberger, L. (1960) *Traité de Botanique Systématique*. Masson et Cie, Paris.

Charlesworth, J.K. (1957) *The Quaternary Era.* Arnold, London.

Coope, G.R. (1962) A Pleistocene fauna with arctic affinities from Fladbury, Worcestershire, *Quart. Journal Geol. Soc. Lond.* 118, pp. 103-23.

Firbas, F. (1949) *Waldgeschichte Mitteleuropas.* Bd. I Jena. Gustav Fischer.

Franks, J.W., Sutcliffe, A.J., Kerney, M.P. and Coope, R. (1959) Haunt of elephant and rhinoceros: the Trafalgar Square of 100,000 years ago.*London Illustrated News* 232, pp. 1011-13.

Godwin, H. (1956) *The History of the British Flora.* Cambridge, Cambridge University Press.

Godwin, H. (1964) Late-Weichselian conditions in south eastern Britain: organic deposits at Colney Heath, Herts. *Proc. Roy. Soc. Lond.* B 160, pp. 258-275.

Hegi, (1966) *Illustrierte Flora der Mitteleuropas,* 2nd. ed.

Hultén, E. (1950) *Atlas of the Distribution of Vascular Plants in N.W. Europe.* Stockholm.

Jessen, K., Anderson, S.T. and Farrington, A. (1959) The interglacial deposit near Gort, Co. Galway, Ireland. *Proc. Roy. Irish Acad.* B 60, 1-78.

McAndrews, J.H. (1966) Post-glacial history of prairie, savanna and forest in northwestern Minnesota. *Torrey Botan. Club, Mem.* 22 (No. 2) 72p.

MacGinitie, H.D. (1958) Climate since the late Cretaceous. pp. 61-79 in *Zoogeography* ed. C.L. Hubbs, Washington D.C.

Meusel, H., Jager, E. and Weinert, E. (1965) *Vergleichende Chorologie der zentraleuropaischen Flora.* Jena, Gustav Fischer.

Oldfield, F. (1968) The Quaternary vegetational history of the French Pays Basque. I. Stratigraphy and Pollen Analysis. *New Phytologist* 67, 677-731.

Pearson, R.G. (1963) Coleopteran associations in the British Isles during the late Quaternary period. *Biological Reviews,* 38, 334.

Puri, G.S. (1950) On a fossil Lotus (*Nelumbo nucifera*) from Kashmir, with a note on the history of the genus. *Indian Forester.* 76, 343.

Puri, G.S. (1951) Fossil fruits of Trapa and remains of other fresh-water plants from the Pleistocene of Kashmir. *Journal of Indian Botanical Society.* 30, 113-121.

Ray, C.E. Wills, D.L. and Palmquist, J.C. (1968) Fossil Musk Oxen of Illinois. *Illinois Academy of Science, Transaction* 61, 282-92.

Rikli, M. (1943) *Das Pflanzenkleid der Mittelmeerländer.* 3 Vols. Berne.

Szafer, W. (1952) *Zarys Ogolnej Geografii Roslin.* Warsaw.

Turrill, W.B. (1929) *The Plant-Life of the Balkan Peninsula.* Oxford.

Walter, H. (1954) *Einführung in die Phytologie.* Bd. III/2 *Arealkunde.* Eugen Ulmer, Stuttgart.

Watts, W.A. (1959) Interglacial deposits at Kilbeg and Newtown, Co. Waterford. *Proc. Roy. Irish Acad.* B 60, 79-134.

Watts, W.A. (1969) A pollen diagram from Mud Lake, Marion County, North-central Florida. *Geological Society of America Bulletin* v. 80, pp. 631-42.

Watts, W.A, and Winter, T.C. (1966) Plant macrofossils from Kirchner Marsh, Minnesota — A paleoecological study. *Geological Society of America Bulletin* v. 77, pp. 1339-1360.

Webb, D.A. (1955) *Erica mackaiana.* Biological Flora of the British Isles. *J. Ecol.* 53, pp. 319-330.

West, R.G. (1957) The Interglacial deposits at Bobbitshole, Ipswich. *Phil. Trans. R. Soc.* B 241, 1-32.

West, R.G., Lambert, C.A. and Sparks, B.W. (1964): Interglacial deposits at Ilford, Essex. *Phil. Trans.* B 247, 185.

West, R.G. and Sparks, B.W. (1960): Coastal interglacial deposits of the English Channel. *Phil. Trans. R. Soc.* B 243, 95-133.

Zeuner, F.E. (1950) *The Pleistocene Period.* Ray Society, London.

CHAPTER II

Intercontinental migration

The migration of plants and animals from one continental land mass to another provides most convincing evidence that land connections once existed where there are none today, e.g. between North America and Eurasia, and between Australasia and south-east Asia. The earliest evidence of migration can also show how long ago those land connections were formed which still exist, for example, between the North American and South American continents.

Several sources of information provide proof of such migrations. The ancestry of living mammals can be traced by fossil lineage and this method proves that migrations have occurred during the evolutionary history of particular groups and also shows which continent was the area of origin and which continent was invaded when the emergence of a land route made migration possible. The geographical location of land connections is best deduced from studying the present distributions of living species that straddle two continents and by considering the occurrence of closely allied plant and animal species which indicate continuity not long ago.

The area in which there are the oldest fossils of a plant or animal group may be taken as the area of origin. Even more certain is the presence of ancestral forms in the fossil record of the area. Once the continent of origin is known, it is possible to say in which direction migration occurred if living representatives or recent fossil remains of the same group occur elsewhere. The completeness of the geological record is important for successful interpretations of this kind. For example, so far as present evidence goes, Africa was the continent in which the Proboscidea (elephants and their allies) originated but it is still possible that central Asia was the area of origin because the late Eocene fauna of warm temperate savanna habitats in that continent is not well known, i.e. the geological record is not complete in this respect (Savage, in Hubbs 1958). Another illustration showing how this method depends on completeness of the fossil record is the history of the Armadillos. Although these are present in the southern United States today, the very complete Eocene fossil record of North America shows no trace of this group at all. It is at least certain, therefore, that this continent was *not* the centre of origin. On this basis it seems likely that the one other continent in which Armadillos are found today (South America) was the area of origin and that the geologically recent appearance of Armadillos in the northern continent represents an immigration.

Fossil ancestors sufficiently like modern mammals to be regarded as belonging in the same family first appear in the middle of the Tertiary era, that is, from the late Eocene to the beginning of the Oligocene period, and thus we need not be concerned with geological history before this. Fossil forms whose affinity with living species places them within the same genera originated between the Miocene and Pleistocene periods while the present-day species themselves evolved entirely within the Pleistocene period. This rapid rate of evolution among mammals contrasts with the longer history of most living plants. However, since each mammal species evolves in a relatively short span of time the period at which migrations occurred from one continent to another can be determined with some precision by reference to mammalian fossils.

Migration between North America and Eurasia

If we compare the fossil record of certain groups in the two continents we find clear evidence of the period and directions in which migration took place. Information from Nebraska is taken to represent mammalian history in North America because of its central position in mid-continent which places it within range of species spreading from both the north-west and the south. It has been in the path of north-south movements during and between glaciations and, above all, it has a very complete series of sedimentary rocks containing mammal bones for all the later Tertiary periods (Tanner and Schultz). The history of different families follows at least three different patterns.

Rhinoceroses have been present on both continents since the Eocene showing that a land connection existed either at that time or before. Significantly, however, they made no appearance in South America and this indicates that no land connections existed here in the early Tertiary. The Eocene rhinos are not represented in Nebraska but Oligocene, Miocene and Pliocene forms are. The rhinoceros line began with a running animal resembling the early ancestors of the horse. It later gave rise to animals of different habitats whose bodily form was modified accordingly, for example water-side animals of massive pig-like proportions. However, the first rhinoceroses to develop a horn on the nose appeared in the Oligocene, e.g. *Subhyracodon,* and were browsing animals with incisors specialised so that they could strip leaves from trees and bushes. From plant evidence already described, we have seen that the American continent was forested from coast to coast at this time and no prairie existed. The two-horned rhinoceroses of Miocene times (*Diceratherium*) took their place and these were also browsing animals feeding on tree foliage. Their Pliocene successors increased still further in size and include the giant American genus *Aphelops,* fourteen feet in length and standing nearly seven feet high. This great size may be related to the grazing habit which developed with the formation of widespread grass and herb vegetation in place of forest on the plains. Then, at the end of the Pliocene, rhinoceroses became extinct in North America and while several rhinoceros species inhabited Europe during the Pleistocene, they had no counterpart on the periglacial plains of mid-west America. The extinction of the American rhinoceros at the end of the Pliocene can only be seen as a failure to produce forms better suited to the changing and increasingly cold climate that developed to extremes of glacial severity in the Pleistocene.

In Eurasia independent evolution succeeded in maintaining a diversity of rhinoceros species in Pleistocene times, some of which were adapted to forest-life (browsers) and flourished during the interglacial periods (e.g. *Dicerorhinus merckii*) and others were suited to the colder tundra environments that existed during the periods of glaciation. The Woolly Rhinoceros (*Coelodonta antiquitatis*) inhabited Europe as recently as the last glaciation and was hunted by Paleolithic man. It also lived in temperate environments as shown by the discovery of a Woolly Rhino skeleton with that of a Straight-tusked Forest Elephant in a deposit dating from the last interglacial at Aveley, Essex. The Woolly Rhino must have co-existed with the more typical "interglacial" mammals in places where a varied landscape brought both grasslands and forest into close proximity.

In the context of intercontinental migration the rhinoceros family history contrasts with others we shall discuss because although it proves the rhinoceros was dispersed throughout the Old World and North America at an early Tertiary date there is no sign that it moved from one continent to the other during its later history. Particularly, the failure of Eurasian Pleistocene rhinos, which included both forest species ("interglacial") and grazing species ("glacial"), to invade the North American continent where endemic species had become extinct does nothing to prove the existence of a land connection but is anomalous when compared with the evidence of other families.

The ancestral horse is represented in Eocene deposits of the Rocky Mountains (not in fact mountainous at the time this animal lived). It was adapted to life in woodlands and to a diet of tree foliage. Through the succeeding geological periods members of the horse family increased in body size and developed longer legs and a tendency to walk on the toes. With diversity in form and especially in complexity of dentition relating to diet, a large number of horses evolved in the Miocene and Pliocene periods. Some remained browsing beasts of various sizes but by the middle Pliocene all of these had become extinct. The increasing dryness of the mid-continent at this time led to a reduction in the cover of woodland, which was replaced by the hard grasses of the earliest prairie. The first grazing horses, whose teeth were suited to this much harsher diet, appeared in the late Miocene (*Merychippus*) and their Pliocene descendants (*Pliohippus* and others) found themselves equipped to exploit a habitat that expanded vastly in area. *Merychippus* moved with its weight supported only on the middle toe of each foot since the others did not reach the ground. *Pliohippus* was the first single-toed horse: the side toes were reduced to long slivers of bone, as in the modern horse. It gave rise to two different types of horses that lived during the Pleistocene. One of these, *Hippidion*, is found (as fossils) in South America and has left no living descendants. The other type, *Equus*, is the genus to which all living horses belong, including Asiatic wild asses (onagers and kiangs), African guagas and zebras.* The Pleistocene provides evidence of the first appearance of horses in the South American continent and these single-toed animals must clearly be derived from predecessors with this same characteristic in the continent to the north. Thus the establishment of a land connection between North and South America occurred no earlier than the late Pliocene. The very limited number of exchanges of fauna that have occurred between these two continents bears witness to the short time that land communication has existed.

The Pleistocene horses of the Old World are all species of the genus *Equus*. This points unquestionably to the existence of a land connection between Eurasia and North America. Remarkably, although horses survived the Pleistocene period in North America they became extinct there only 8,000 years ago, possibly exterminated by paleolithic hunters. There were no wild horses in America when the first Europeans arrived and those that were used subsequently by the native Indians escaped (or were captured) from stocks introduced by the Spaniards.

The camel family is not a group one would associate with North America today but its fossil record shows that it was in fact the original home of the camels and here they evolved from the early part of the Oligocene up to the Pleistocene, becoming extinct only about 8,000 years ago. Although they have a long and complete history in Nebraska,

* The wild ass of Asia is a widely distributed animal of the plains and is known by different common names in its regionally distinct races e.g. the Kulan in Mongolia, the Kiang of the high plateau of Tibet, and the onager or Persian wild ass. Zoologically these are considered to be conspecific and all are named *Equus hemionus*. Przewalsky's Horse is the other surviving Eurasian wild horse, possibly still to be found native in the Gobi-Altai region of Mongolia (Montagu, 1968) and maintained in captivity in Europe.

camel species of today occur in continents where there are no pre-Pleistocene fossils of this kind. Outside the U.S.A. they appear suddenly in the fossil record at this geologically recent time. There is nothing on the skeletons of either the modern or the extinct forms to indicate whether the animals had humps or not. The llama and its wild relatives the guanaco and vicuña of south America are as truly camels as the Dromedary (one-humped camel) of Arabia and the Bactrian (two-humped camel) of central Asia. The Bactrian camel is well adapted for life in cold deserts where snow and severe low temperatures are experienced in winter. It has hair a foot long and in this respect resembles several now extinct mammals of the Pleistocene glacials.

From their origin as small animals the camels evolved in Nebraska and adjacent areas to species of giant size in the early Pleistocene. *Gigantocamelus fricki* attained a height of eleven feet. This spectacular increase in size culminating in the Pleistocene forms has been noted also in other families. The extinct camels of the American mid-west are usually found in deposits from interglacial periods in which they are associated with a southern type of fauna that indicates warmer conditions. This fauna immigrated from the south at the onset of each interglacial and retreated in ensuing glaciations. By the early Pleistocene camel species had established themselves in South America, which shows that a land route had come into being by this time. As in the case of horses, the entry of camels into Asia during the Pleistocene proves that this continent was joined to North America and in both examples it is remarkable that the continent in which surviving species now flourish was so recently invaded by their Pleistocene ancestors.

There is evidence of migration from Eurasia to North America in the fossil record of other families, notably elephants (*Proboscidea*) and bovines. Neither of these groups existed in North America up to the Pleistocene and had originated and evolved in the Old World. Indeed their first appearance is in deposits later than the Kansan (i.e. second) glaciation. Their first representatives in North America are cold-climate species including Mammoths (*Mammuthus*), Musk Oxen (*Symbos cavifrons* and *Ovibos moschatus*), Bison (*Bison latifrons*) and wild cattle (*Platycerabos*). It strongly suggests that they immigrated sometime during the glaciation and in the following one. Once they were established within the continent elephants evolved endemic species adapted to a temperate climate. In fact the world's largest elephants, such as *Archidiskodon maiberi* and Columbian Mammoth (*Mammuthus columbi*), occurred in the interglacial periods of America, at the same time as the Straight-tusked Forest Elephant, *Elephas antiquus,* in Europe.

The first bison in North America like their Eurasian predecessors had an immense horn span of more than eight feet but they become successively smaller up to the present. Other bovine animals were large and while in North America none except the bison survived after the Pleistocene, among the living species of Asia the Yak (*Bos grunniens*) is probably a little-altered descendant of the Pleistocene cattle. Indeed the environment in which the Yak has survived in Tibet, a plateau over 14,000 feet in elevation which experiences temperatures as low as − 40°C, is a vestige of conditions that were extensive over all mid continent lowlands during the glacial period.

The similarity of the fauna of North America and Eurasia in the Pleistocene period was more marked than today because extinction has removed some elements from one continent or the other seemingly at random. However, the similarity that did develop was in very striking contrast to the distinct faunas of these continents throughout the Tertiary. The sudden interchange of faunas that occurred shows beyond doubt that a land route connected the land-masses of the northern hemisphere at least intermittently during the Pleistocene.

There is equally clear evidence that a small contingent of mammals that evolved in South America immigrated into the southern part of North America, penetrating as far north as Nebraska during the interglacial phases. These animals include Armadillo, Peccary, Tapirs and Ground Sloth. The last two now inhabit sub-tropical forests and after their interglacial excursions into North America became extinct there. The forested land route between North America and South America that exists today thus appeared for the first time in the Pliocene. Of course, migrations may take place on successive occasions by the same land route if its existence is alternately broken and re-established by eustatic changes in world sea-level. As we have seen, migration may occur in both directions, possibly but not necessarily at the same time. There is no need to imagine coincident migration in opposite directions unless animals of similar habitats and climatic tolerance are represented, e.g. large herbivores of cool temperate prairie or steppe. In all cases, what we call migration is in fact a gradual movement of populations extending their grazing grounds generation by generation and thus occupying the recently emerged land surface as it is colonized by plants and clothed with a vegetation. Thus animal migration and plant migration are intimately connected and the "crossing of land bridges" between continents is the gradual occupation of new land surface by the developing ecosystems suited to the prevailing climate of the time. The central American land-bridge probably provided conditions suitable for the southward migration of camels during the glacial period at the higher elevations across the isthmus, and quite different conditions suitable for the northward migration of sloths and tapirs during interglacial periods at low elevations.

Modern distributions as evidence of migration

Even if no fossil remains were available the pattern of distribution of many living species in the holarctic region is sufficient evidence that a land connection existed between the northern continents at a very recent period in geological time. The argument is that where plants and animals of one continent resemble their counterparts on another continent so closely that they cannot certainly be distinguished then the present isolation of the two groups cannot have been long established. Isolation maintained for long periods of time generally results in the development of differences sufficiently marked for the individuals from different areas to be identified by separate names. Closely related species, i.e. forms that are distinct but still fairly similar, are held to indicate relatively recent continuity while more distantly related species, those with greater dissimilarities, indicate continuity at an earlier time followed by a longer period of isolation.

A number of animals of North America and Eurasia belong to the same species and it is informative to note exactly which species and what kind of animals these are. They include the polar bear, arctic fox, arctic hare, grey wolf, wolverine, two species of weasels, two of voles, a shrew and a ground squirrel. The arctic and sub-arctic character of this fauna is unmistakable. Add to these a further list of closely related species found on the two continents, some of which are considered merely separate geographical races of the same species, and the affinity is obviously extremely close. The European and American names are given as well as the scientific ones in Table 9.

Table 9   **Representative mammals of the holarctic fauna.**

Common name in N. America	in Eurasia	Scientific name	Family
Moose	Elk	Alces alces A.a. americana	Cervidae (Deer)
Caribou	Reindeer	Rangifer tarandus R. arcticus	Cervidae (Deer)
Wapiti	Red Deer	Cervus elaphus C. canadensis	Cervidae (Deer)
Bison	Bison or Wisent	Bison bonasus B. bison	Bovidae (Cattle)
Brown Bear (Alaskan)	Brown Bear (European)	Ursus arctos	Ursidae (Bears)
Red Fox	Red Fox	Vulpes vulpes V. fulva	Canidae (Dogs)
Fisher	Marten	Martes martes M. pennanti	Mustelidae (Stoats)
Otter	Otter	Lutra lutra L. canadensis	Mustelidae (Stoats)
Beaver	Beaver	Castor fiber C. canadensis	Castoridae
Lynx	Lynx	Lynx lynx L. canadensis	Felidae (Cats)
Bob Cat	—	Lynx rufus	Felidae (Cats)
	Wild Cat	Felis sylvestris	Felidae (Cats)

This assemblage of mammals, in which a small degree of differentiation is recognised between the North American and Eurasian equivalents, includes species whose distribution has great range in latitude and while some of them may seasonally cover arctic territories all are inhabitants of boreal forest, woodland or prairie. If we consider, in the case of arctic mammals that some interchange of individuals can take place across the polar ice, we cannot suggest the same is true of the boreal mammals. The distinctions that have developed between Eurasian and North American populations indicate that they are effectively isolated from each other under present conditions and have been so for some time. When communication existed there would presumably have been land connection within boreal latitudes and vegetation cover appropriate to these mammals, especially the herbivorous species.

The distribution of living plant species in the northern latitudes of both continents also provides evidence for some former land connection between them and points strongly to the Bering Straits as the site of this link. Hultén has discussed in detail the areal groups into which about 2,000 plants of this region can be arranged on the basis of their total distributions, i.e. each considered in its entirety. He finds that the flora of the area east of the Lena River and west of the Mackenzie River does not reveal very significant interruption at the Bering Strait. The plant-life of the two continents is strikingly similar in these latitudes: even the same species are present in both the east Siberian and Alaskan territories. Many of them have distributions of horseshoe form centred around the Straits and extending along the two limbs, eastward into North America and westward into Eurasia, to various distances in different species. Thus, far from being a dividing line the Bering Straits appear to have functioned as a bridge and even as a centre of plant migration, despite the fact that today they are a natural sea barrier. The kinds of plants represented include an oceanic group, an arctic group, arctic-montane and boreal groups, each of which diverges from the Straits region in a characteristic direction, i.e. along the north Pacific coastlands, along the coasts of the Arctic Sea, along the mountain ranges and across the lowland interior of the continents. Species such as *Cassiope lycopodioides* and *Geum rotundifolium* which extend from Washington and Oregon through Alaska to the Bering Sea and through the Aleutian Islands to Kamchatka and Hokkaido, convey very strongly the impression of the inter-continental link. Some less extensive species occupy just the middle part of this chain, e.g. the Woodrush (Fig. 11.1). Examples of this oceanic group might suggest that the connection is nothing more than "island-hopping" but this can be refuted. The Aleutian Islands were heavily glaciated and were probably occupied by plants quite late in Pleistocene history as climatic warming caused wastage of the glaciers and as sea-level rose, when they

LUZULA
● PIPERI
○ WAHLENBERGII

11.1   The trans-Beringian distribution of the woodrushes, (a) *Luzula wahlenbergii*
(b) *Luzula piperi* (after L. Hamet-Ahti, *Aquilo*, 1965).

became island refuges, i.e. remnants of the once extensive land
connection.

A much larger group is that of arctic-montane plants which are
present in the tundra territories bordering the Arctic Sea and also
extend (with interruptions) along the mountain chains, south-east along
the Rockies and south-westward into central Asia to the Altai

Potentilla fruticosa L.

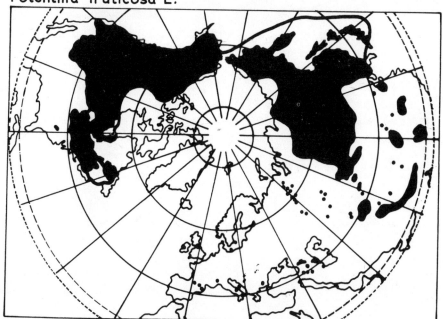

11.2   The semi-circumpolar distribution of Shrubby Cinquefoil (*Potentilla
fruticosa*). Note its continuity in the Beringian sector and its relict status in
W. Siberia and Europe. (After Meusel *et al* 1965).

Mountains. Trailing Azalea and Alpine Clubmoss are examples of this distribution and also occur in the Aleutians, which is unusual in the group. Most arctic-montane plants show the linkage by their presence on both sides of the Bering Strait but not in the Aleutians. Among the most continuous and widespread of such types are Mountain Avens, Mountain Sorrel and Viviparous Knotgrass, which are almost completely circumpolar and are present, for example, on mountains in the British Isles. Other arctic species have gaps in their circumpolar distribution, which therefore has the form of a horseshoe not a ring, but significantly the gap never coincides with the Bering Strait! Here we may refer to Moss Campion (gap east of the Ural Mountains), Opposite-leaved Saxifrage (gap in the Rocky Mountains), Alpine Meadow-rue and Three-flowered Rush (gap in arctic America).

Those plants termed "boreal" have a much broader distribution in middle latitudes than the truly arctic. They include plants of heaths (which may extend into the arctic) and northern forests. Some species characteristic of both these types of vegetation have extensive circumpolar distributions. Examples are *Empetrum nigrum*, *Vaccinium uliginosum* and *Vaccinium vitis-idaea* (heaths), and the Wintergreens (*Pyrola rotundifolia* and *Pyrola secunda*) which inhabit damp rock ledges and woods. Other boreal species have the horseshoe pattern of

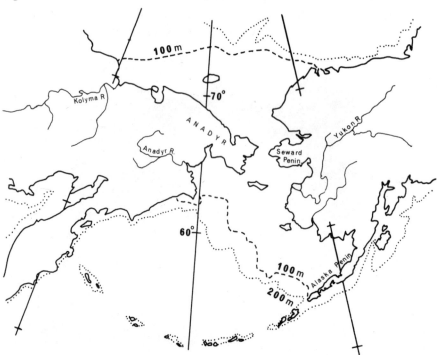

11.3 Map of the Bering Strait region showing the probable extent of Beringia (100m submarine contour) during the glacial stages of the Pleistocene.

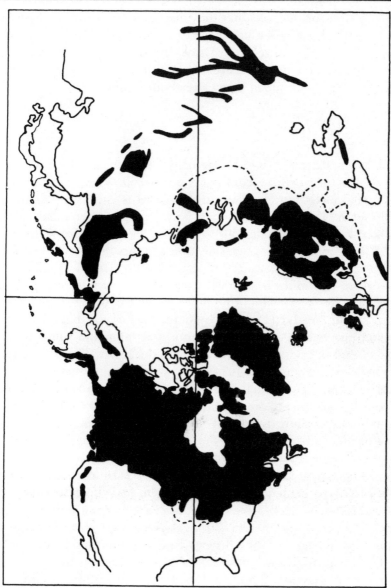

11.4 The extent of ice-cover in the northern hemisphere (black) during the last glacial period. Also shown (broken line) is the boundary of ice-cover in earlier glacials where this was more extensive. (After Hadac, in Löve & Löve 1963).

distribution with more or less limited westward extension in Eurasia, e.g. Shrubby Cinquefoil (Fig. 11.2), (continuous to Yenisei R., relict further west), Leatherleaf (reaching eastern Sweden) and Andromeda (reaching Ireland). However, in all these cases continuity of distribution across the Bering Strait makes the former existence of a land connection certain. We may gain some idea of the conditions that existed on the land-link from the ecological character of the plants which exist in that region now, and also from those whose bi-continental distribution suggests that they migrated via this route. Species of tundra, heath and bog are in a majority, a small number of forest herbs are represented but no species of trees are shared by both continents. The ecological indications accord well with those we can infer from the character of the widespread fauna already discussed.

The Bering land corridor

During every glaciation the storage of water in the great ice-sheets that formed over the northern continents resulted in a lowering of sea-level and for the maximum glaciation the sea-level is estimated to have sunk by 90-100 m. The continental shelf between east Siberia and Alaska is unbroken between latitudes 60° and 75°N, extending from the tip of the Alaska Peninsula to a considerable distance north of the modern coastline of the Arctic Sea. A fall in sea-level of 100 m would expose a land surface 800 miles from north to south and even lowering by 50 m would provide a corridor 250 miles in width between the Seward Peninsula and Anadyr (Fig. 11.3). There is no doubt that a land-bridge of something between these maximum and minimum dimensions has been exposed and submerged several times during the Pleistocene period as glaciations and interglacials have followed one another. Geomorphological evidence shows that large areas of eastern Siberia were never covered by ice-sheets and that glaciation was confined to the mountains. The same is true of those parts of Alaska bordering the Bering Strait, and the valley of the Yukon River remained ice-free, like its Siberian counterpart the Anadyr (Fig. 11.4).

The connecting land surface and its northern extension into the Arctic Sea were named Beringia by Hultén, and these areas were also presumably unglaciated. Even during glacial times the southern border of Beringia from the Alaska Peninsula westward was subject to the oceanic influence of the Bering Sea and its moderate latitude (55°-60°N) ensured a warm season of at least four to five months. As the northern margin of Spruce forest lies only 50 km south of the Seward Peninsula at the present time, i.e. almost 65°N, we can imagine that such forest survived in close proximity to the Asian-Alaskan land corridor even if forest trees did not themselves colonize it. There is thus very good correspondence between the evidence of fauna, flora and geomorphology. Furthermore, as a result of recent studies by Colinvaux of the sub-fossil record in Alaskan lake sediments the Pleistocene history of the region is confirmed.

The North Atlantic problem

The possibility of a series of land connections linking North America,

Greenland, Iceland, Faroes, the British Isles and Europe has been argued by some authors, notably Lindroth (1963) and Dahl (1963). They attempt to account for the peculiar distribution of some species whose areas are appropriately called amphi-atlantic, including examples among both plants and beetles. They are generally arctic or boreal species but unlike those which are circumpolar and those which are entirely North American or Eurasian, the amphi-atlantic species occupy relatively limited areas near to the Atlantic margins of both continents and usually also the islands between them in the North Atlantic. Their distribution differs from that of Beringian species in that the areas of territory they occupy are small in relation to the distance that separates them. As an example the distribution of one of the arctic-montane plants, *Potentilla crantzii,* is shown (Fig. 11.5). Hultén maintains that such geographical types are the ultimate stage in the reduction of formerly

Potentilla crantzil (C.F.) BECK

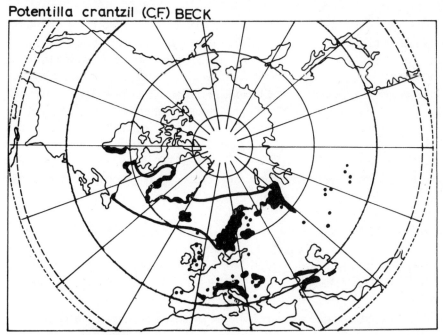

11.5 The amphi-atlantic distribution of the Cinquefoil, *Potentilla crantzii,* (after Meusel *et al* 1965).

wide or even circumpolar distributions. He presents a series of maps for various species which illustrate intermediate conditions in such a process, i.e. some having disjunct – and obviously relict – stations in western America (*Tofieldia palustris*) or in eastern or central Asia (*Juncus stygius, Juncus trifidus*).

Of course, lowering of sea-level during the glacial periods would be effective here too in exposing the continental shelf. This, however, extends the European mainland principally in the North Sea and Bay of Biscay. The Scandinavian coast would be extended only to the Lofoten Islands and Scotland would be linked with the Shetlands, but deep and wide sea barriers would still remain in the Davis Strait between Baffin Island and Greenland and in the Denmark Strait between Greenland and Iceland. Reduction of sea-level by 200 m would not alter this situation and the extrusion of Tertiary lavas along a belt extending from Iceland through the Hebrides to Northern Ireland is not believed to have modified the bathymetry. Thus geological evidence in this area is strongly against the proposition that any land link has existed within later Tertiary times or more recently than this (Einarsson, 1964).

The Central American link

When we discussed the Pleistocene mammals of North America we referred to South American groups entering the continent during the Pleistocene. Only at this time did the antecedants of the llama and vicuña (Camel family) reach the southern continent from the northern. The Central American land corridor was formed by tectonic movement in the late Pliocene and early Pleistocene at the same time as Cordilleran orogeny in the Andes and the Sierra of California. The Pleistocene fossil fauna of Florida includes the earliest known records in North America of several mammalian families whose earlier history and evolution had taken place in South America. These include types which subsequently spread northward and now have living members in the northern continent, e.g. Porcupine (*Erethizon*), Armadillo (*Dasypus*), Peccary (*Tayassu*), and others that flourished during the interglacials but later became extinct in North America. Among the latter should be mentioned the Ground Sloths, *Megatherium* (in Florida), *Nothotherium* (Texas) and *Paramylodon* (known from fossils in Texas, Florida, Kansas, Colorado and Nebraska) and that other forest-dweller the Tapir which has left fossil remains in north-central and south-west Oklahoma and at El Paso and the Hueco Mountains in Texas. These distributions imply that forest ecosystems were extensive during the interglacial periods (or during parts of them at least) in many states that are occupied by prairie today. It is thus apparent that grasslands have not continuously occupied their present areas through the Pleistocene. Another clue pointing to the existence of moister interglacial conditions in the mid-west is the fossil occurrence in central Kansas of the Water Rat (*Neofiber*) which is now limited to swamp situations in the Florida peninsula. Florida itself has evidence of other aquatic invaders from South America in the fossil remains of the Capybaras (Rodent family). Other living species whose relatives are mainly in South America are the marsupial Opossum and the Raccoon (*Procyon*) which by their relationships show that migrations across the Panama corridor have taken place.

The physiography and climate of the Central American land link provide both forested lowland and temperate elevated routes for migration but it is interesting that the ever-wet forest zone is much narrower than the land-mass itself and the mountain chain is inter- rupted by several lowland troughs (Fig. 11.6). The only continuous south-to-north zone that experiences rainfall all the year round at low elevations lies on the eastern or Caribbean flank of the mountain spine and is restricted to a corridor less than 50 km wide in northern Honduras. Pleistocene lowering of sea-level can have broadened this corridor but slightly. Only in this zone, where annual rainfall exceeds 2000 mm and is distributed through a humid period of 10-12 months, does evergreen wet tropical forest extend uninterrupted between North and South America. Plants and animals of the hot, wet tropics of South America extended northward principally in this zone while those tolerant of a pronounced dry season of 5-6 months followed the equally narrow west coast (Pacific) route to the Sonoran region of North America. One of the low points in the land link, the Nicaragua depression, was marine until the Miocene period and even after its emergence it underwent an extensive transgression from the Pacific Ocean as late as the upper Pliocene. The lowland connection therefore existed first during the Pliocene and again during the Pleistocene.

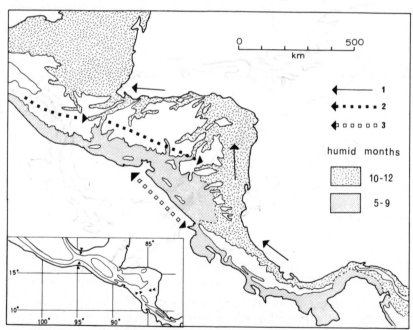

11.6   Climatic-orographic divisions of Central America (after Lauer) and migration routes. Caribbean lowlands have a highly humid climate; Pacific lowlands experience a dry season. Highlands above 800m (shown white) have a temperate climate. (1) Route of migration for forest-dwelling tropical species, (2) route for temperate and montane species, (3) route for species of seasonally arid environments. Inset: the interruption of highland routes at the Isthmus of Tehuantepec and the Nicaragua depression, marked ▶▶◀◀ , with generalised contours of 500m and 1500m (after Hastenrath).

The present distribution of South American and North American elements on the land link generally shows marked altitudinal segregation, the former being lowland and the latter montane. For example, at elevations above 1000 m trees of the selva are left behind and a zone of oak-pine forests is entered. This is true for mountains as far southward as south Nicaragua but here many North American plants have their southern limit. Pines, for example, which have a centre of abundance and many species in the Mexican meseta are absent from South America and do not occur in the mountains of Costa Rica (i.e. south of the Nicaragua depression). Conversely, many high montane plants of the southern continent reach their northern limit in Costa Rica. A third pattern is illustrated by oaks (*Quercus*) which at temperate elevations occur on all Cordilleran ranges southward to Colombia.

The barrier to migration presented by the gaps in the mountain chain are most formidable to the flora and fauna of cool-temperate and high montane environments since the distances between such locations are greater than in the case of habitats found at lower elevations. Thus, physiographic and ecological conditions on an intercontinental land link create a "filter effect" in which the opportunities for migration are not equal for all biota. These conditions tend to interpose, as it were, a sieve of variable mesh through which plants and animals of differing environments may pass with more or less difficulty. The Pleistocene climatic changes of the northern hemisphere seem to have had minimal effect in Central America where the snow-line was lower by some few hundred metres (to about 3,500 m). Since there are no elevations exceeding 3,000 m between Costa Rica and the Guatemalan Highlands the opportunities for migration of montane species can have been improved only marginally at that time. On the other hand the continuity of forest environments at all levels would not have been affected.

South-east Asia

The distribution of species common to the Malaysian and Indonesian islands and the continent of south-east Asia demonstrate that land connections once existed between them. The wet forests of Sumatra, Java and Borneo are similar to those of the Malay peninsula not only in general appearance and physiognomy but also in the close relationship of the plants that comprise them and the animals that inhabit them. The continental shelf extends to the Macassar Strait, east of Borneo, and much of it must have been exposed in the area of the South China Sea as the Sunda Shelf when world sea-levels were lower during the Pleistocene (Fig. 11.7). Pines reach their southern limit here in the Old

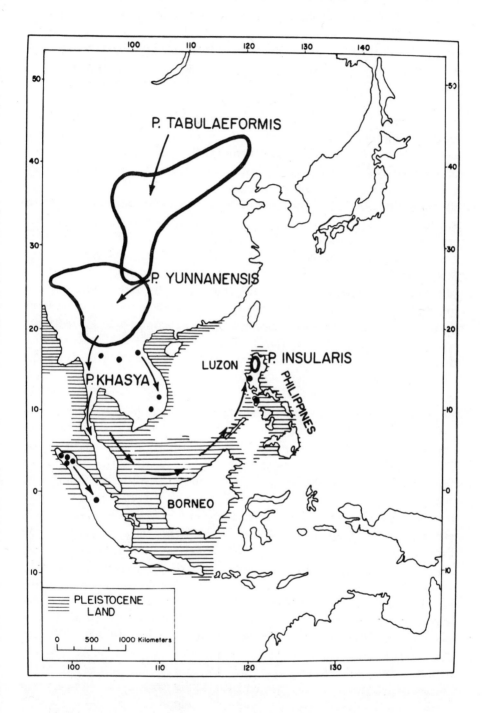

11.7 The robable extent of land connections in south-east Asia during glacial
periods of the Pleistocene. Also shown is the modern distribution of certain
pines and the presumed routes of their migration. (From N.T. Nirov, *The
Genus Pinus,*© 1967 Ronald Press Company, New York.

World as they did on the Central American link in the New World. The distribution of various pine species suggests that the path of migration swept southward from China in a great arc, leaving relict areas of *Pinus khasya* in Sumatra, and probably reached the Philippines via Borneo. The rift between Formosa and the Philippines was formed in the early Tertiary so that no terrestrial link between these can have existed since then.

The islands situated near the margin of the Sunda Shelf, Java, Borneo and the Philippines possess a minority element in their flora and fauna derived from the Australasian continent, the margin of which extends to western New Guinea as the submarine Sahul Shelf. Some parallels with the situation in Central America can be detected. Species related to those of the Asian and Australasian continents are segregated by altitude and climate in the islands. The mountain flora of the islands generally contains most of the plants that are Australasian, certainly in New Guinea and Borneo (but not so in Java). The lowland evergreen forest of Borneo is composed almost exclusively of Indo-Malaysian plants but at elevations above 1,000 m there are montane forests of quite different composition in which some of the commonest trees are Australasian e.g. *Weinmannia, Dacrydium* and *Phyllocladus.* On Mount Kinabalu, the highest mountain in Borneo (13,455 feet), some exciting discoveries were made by the Royal Society Expedition of 1961 and this can be ascribed to the fact that its upper slopes form an island of climatically temperate environments in a vast sea of lowland forest. The mountains of Java (and Sumatra) do not include this Australasian element and the mossy forest or cloud forest in west Java is composed chiefly of trees of *Vaccinium* which is distinctly of Eurasiatic origin. This is analogous with the occurrence of Tree Heath (*Erica arborea*) on the East African peaks. In Java a different Australasian component occurs including *Eucalyptus* and *Casuarina* and the factor that is responsible for its segregation from the ever-wet forests is seasonality of climate rather than altitudinal climate. Here an analogy with the Pacific coastal lowland of the Central American link is apparent. From east Java, through the Lesser Sunda islands to northern Australia a strongly seasonal monsoon climate prevails and tolerance of a dry season is a requirement in species that grow here as well as in those that may have migrated by former land connections in this zone. Indeed, Zeuner suggested that two separate routes with differing climatic characteristics have at some time linked the Malayan and Australasian regions, one with year-round rainfall and dense evergreen forest connecting Borneo via Celebes and the Moluccas to New Guinea, while further south seasonally dry climate dominated any possible link with Java from the east. All the complexities of filter effects and varied landscape characteristics (e.g. landforms and soil types) must have made the history of migration in this part of the world an intricate one.

Moreover, we are still waiting for geophysical evidence for an emergent connection between the Sunda and Sahul Shelves. Thus while the orogenic activity of a major volcanic belt has undoubtedly played an essential role here, the Pleistocene lowering of sea-level and exposure of a very large area of continental shelf can alone account for much of the biogeography of the south-east Asian archipelago.

SOURCES OF REFERENCE

de Beaufort, L.F. (1954) *Zoogeography of the Land and Inland Waters.* London, Sidgwick and Jackson.
Blair, W.F. (1958) Distributional patterns of vertebrates in the southern United States in relation to past and present environments. pp. 433-468 in *Zoogeography* ed. C.L. Hubbs.
Burt, W.H. (1958) History and affinities of the recent land mammals of western North America, pp. 131-154 in *Zoogeography* ed. by C.L. Hubbs.
Corner, E.J.H. (1964) Royal Society expedition to North Borneo 1961: Reports. *Proc. Linn. Soc. Lond.* 175, 9-56.
Dahl, E. (1963) Plant migrations across the North Atlantic Ocean and their importance for the paleogeography of the region. pp. 172-188 in *North Atlantic Biota* ed. A. and D. Löve.
Einarsson, T. (1964) On the question of Late-Tertiary or Quaternary land connections across the North Atlantic, and the dispersal of biota in that area. *Journal of Ecology*, 52, 617-625.
Hansen, H.A., ed. (1967) *Arctic Biology.* Oregon State University Press, Corvallis.
Hastenrath, S. (1968) Certain aspects of the three-dimensional distribution of climate and vegetation belts in the mountains of Central America and Southern Mexico, in *Colloquium Geographicum* Bd 9, 122-30. Bonn.
Heezen, B.C. and Tharp, M. (1963) The Atlantic Floor in *North Atlantic Biota* ed. Löve, A. and Löve, D.
Hultén, E. (1937) *Outline of the History of Arctic and Boreal Biota During the Quaternary period.* Stockholm
Hultén, E. (1958) *The Amphi-atlantic Plants and their Phytogeographical Connections.* Almquist and Wiksell, Stockholm.
Hultén, E. (1964) *The Circumpolar Plants.* Almquist and Wiksell, Stockholm.
Hultén, E. (1963) Phytogeographical connections of the North Atlantic. in *North Atlantic Biota*, ed. by A. and D. Löve.
Lauer, W. (1968) Problemas de la division fitogeografica en America Central. *Colloquium Geographicum* Bd 9, 139-156. Bonn. Proceedings of a symposium, "Geo-ecology of the mountainous regions of the Tropical Americas", ed. C. Troll.
Lindroth, C.H. (1963) The problem of late land connections in the North Atlantic area. pp. 73-85 in *North Atlantic Biota*, ed. Löve, A. and Löve, D.
Montagu, I. (1968) Przewalski Horses in the wild. *Oryx. 9*, 260-2.
Pearson, R.G. (1964) *Animals and Plants of the Cenozoic Era.* Butterworths, London.
Richards, P.W. (1943) The ecological segregation of the Indo-Malayan and Australian elements in the vegetation of Borneo. *Proc. Linn. Soc. Lond.* 154, pp. 154-6.
Savage, D.E. (1958) Evidence from fossil land mammals on the origin and affinities of the western Nearctic fauna. pp. 97-129 in *Zoogeography* ed. C.L. Hubbs.
Schultz, C.B. and Tanner, L.G. Numerous articles on Tertiary and Pleistocene fauna in: University of Nebraska, *Museum Notes*, Nos. 8, 11, 12, 15, 18, 19. (1959-62). Lincoln, Nebraska.
Shaw, H.K.A. (1943) The biogeographic division of the Indo—Australian Archipelago: 5. Some general considerations from the botanical standpoint. *Proc. Linn. Soc. Lond.* 154, pp. 148-54.
Zeuner, F.E. (1943) The biogeographic division of the Indo-Australian Archipelago. 7. *Proc. Linn. Soc. Lond.* 154, 157.

Further Reading

Exploration and survey

Ritchie, J.C. (1962) *A Geobotanical Survey of Northern Manitoba.* Arctic Institute of North America, Technical paper 9.
van Steenis, C.G.J. (1950) Desiderata for future exploration. *Flora Malesiana.* Ser. 1, vol. 1, p. cxii Djakarta.

Geographical elements

Dice, L.R. (1943) *The Biotic Provinces of North America.* University of Michigan Press. Ann Arbor.
Gleason, H.A. and Cronquist, A. (1964) *The Natural Geography of Plants.* Columbia University Press, New York & London.
Matthews, J.R. (1955) *Origin and Distribution of the British Flora.* Hutchinson, London.
Strasburger's Textbook of Botany, translated by Bell, P. & Coombe, D. (1965) Longmans, London, pp. 751-86.

Influence of local factors

Evans, E. Price (1945) The study of the distribution of floristically rich localities in relation to bedrock. *J. Ecol.* 32, pp. 167-179.
Lindroth, C.H. (no date) Skaftafell, Iceland.*Oikos* Supplement No. 6 (on relation of beetle faunas to microclimate).
Taylor, J.A. (1967) Soil climate: its definition and measurement. pp. 37-47 in *Weather and Agriculture,* ed. Taylor. Pergamon, London and New York.

Plants and climate

Conolly, A.P. & Dahl, E. (1970) Maximum summer temperature in relation to the modern and Quaternary distributions of certain arctic-montane species in the British Isles. pp. 159-223 in: Walker, D. & West, R.G. (eds.) *Vegetational History of the British Isles.* Cambridge, Cambridge University Press.
Faegri, K. (1950) On the value of palaeoclimatological evidence. *Centenary Proceedings Roy. Met. Soc.* pp. 188-195.
Faegri, K. (1958) On the climatic demands of oceanic plants. *Botaniska Notiser* 111, pp. 325-332.
McVean, D.N. & Ratcliffe, D.A. (1962) Climate and vegetation. Chapter 11 in *Plant Communities of the Scottish Highlands.* H.M.S.O., London.
Ouellet, C.E. & Sherk, L.C. (1967) Woody Ornamental Plant Zonation: I. Indices of winterhardiness. pp. 231-8, II. Suitability indices of localities. pp. 339-49, III. Suitability map for the probable winter survival of ornamental trees and shrubs. pp. 351-8. *Can. J. Plant Sci.* 47.
Taylor, J.A. (1967) Growing season as affected by land aspect and soil texture. pp. 15-36 in *Weather and Agriculture* ed. J.A. Taylor. Pergamon, London and New York.

Plant life forms

Clapham, A.R. (1935) Review of *The Life Forms of Plants and Statistical Plant Geography.* (The collected essays of Carl Raunkiaer) *J. Ecol.* 23, pp. 247-9.
Dansereau, P. (1951) Description and recording of vegetation upon a structural basis. *Ecology* 32, 172.
Du Rietz, G.E. (1931) Life-forms of terrestrial flowering plants. I *Acta Phytogeographica Suecica* 3.
Ellenberg, H. & Mueller-Dombois (1967) A key to Raunkiaer life-forms with revised subdivisions.*Veröff. Geobot. Forschungsinstitut Rübel* 37, pp. 56-73 Zürich.
Smith, W.G. (1913) Raukiaer's life-forms and statistical methods. *J. Ecol.* 1, pp. 16-26.

Classifying vegetation

Beard, J.S. (1969) The natural regions of the deserts of Western Australia. *J. Ecol.* 57 (3), 677-712.
Dansereau, P. & Arros, J. (1959) Essais d'application de la dimension structurale en phytosociologie. *Vegetatio* IX, pp. 48-99.
Küchler A.W. (1949) A physiognomic classification of vegetation. *Annals of the Association of American Geographers* vol. 39, 201-210.
Laubenfels, D.J. de (1968) The variation of vegetation from place to place. *Professional Geography* 20, pp. 107-111.
Ross Cochrane, G. (1963) — includes a synopsis of Küchler's classification in: A physiognomic vegetation map of Australia. *J. Ecol.* 51 (3) 639-55.
Tansley, A.G. (1913) A universal classification of plant communities. (A review of the system of Brockmann-Jerosch and Rubel, 1912) *J. Ecol.* 1, pp. 27-42.
Various authors (1955) Colloques sur les régions écologiques du globe. *Année biologique* 3e Série, t. 31, fasc. 5-6. Also published by Centre Nationale de la Recherche Scientifique as *Les divisions écologiques du monde: moyens d'expression, nomenclature, cartographie.* Paris.

Migration and dispersal

Cain, S.A. (1944) *Foundations of Plant Geography.* Harper, New York.
Elton, C.S. (1958) *The Ecology of Invasions by Animals and Plants.* Methuen, London.
Gleason, H.A. & Cronquist, A. (1964) *The Natural Geography of Plants.* Columbia University Press, New York.
Johnson, C.G. (1969) *Migration and Dispersal of Insects by Flight*, 763 pp. Methuen, London.
Löve, Doris (1963) Dispersal and survival of plants. pp. 189-205 in: *North Atlantic Biota and their History.* Ed. A. & D. Löve. Pergamon, London and New York.
Webb, D.A. (1966) Dispersal and establishment: what do we really know? pp. 93-102 in *Reproductive Biology and Taxonomy of Vascular Plants.* (ed. J.G. Hawkes) Pergamon, Oxford.

Relicts and refugia

Allee, W.C. & Schmidt, K.P. (1951) *Ecological Animal Geography.* 2nd edn. Wiley, New York; Chapman & Hall, London.
Holmquist, C. (1962) The relict concept. *Oikos* 13, 262-92.
Pearson, R.G. (1965) Problems of post-glacial refugia. *Proc. Roy. Soc. Lond.* B 161, pp. 324-30.
Piggot, C.D. & Walters, S.M. (1954) On the interpretation of the discontinuous distribution shown by certain British species of open habitats. *J. Ecol.* 42, pp. 95-116.

Changes in distribution

Butzer, K.W. (1965) *Environment and Archaeology: An Introduction to Pleistocene Geography*. Methuen, London.
Frenzel, B. (1968) The Pleistocene vegetation of Northern Eurasia. *Science*. 161, No. 3842, pp. 637-649.
Hoffman, R.S. & Taber, R.D. (1968) Origin and history of holarctic tundra ecosystems, with special reference to their vertebrate faunas. pp. 143-170 in: *Arctic and Alpine Environments*, Wright, H.E. & Osburn, W.H. (ed.) Indiana University Press, Bloomington.
Mitchell, G.F. & Watts, W.A. (1970) The history of the Ericaceae in Ireland during the Quaternary epoch. pp. 13-21 in: Walker D. & West, R.G. (ed.) 1970. *Vegetational History of the British Isles*. Cambridge.
Sculthorpe, C.D. (1967) *The Biology of Aquatic Vascular Plants*. Arnold, London. Chapter 11: Some aspects of the geography of aquatic vascular plants. pp. 365-413.
Shotton, F.W. (1965) Movements of insect populations in the British Pleistocene. *The Geological Society of America. Special paper* 84, pp. 17-33.
Smith, P.W. (1957) An analysis of post-Wisconsin biogeography of the Prairie Peninsula region based on distributional phenomena among terrestrial vertebrate populations. *Ecology* 38, pp. 205-218.

Intercontinental migration

(a)   Beringia
Colinvaux, P.A. (1964) The environment of the Bering land bridge. *Ecological Monographs* 34, p. 297.
Haag, W.G. (1962) The Bering Strait Land Bridge. *Scientific American* 206, p. 112.
Hopkins, D.M. (ed.) (1967) *The Bering Land Bridge*. Stanford, Stanford University Press.

(b)   The North Atlantic
Lindroth, C.H. (1957) *The Faunal Connections Between Europe and North America*. Almquist & Wiksell. Stockholm.
Löve, A. & Löve, D. (ed.) (1963) *North Atlantic Biota and their History*. Pergamon, London and New York.

(c)   Central America
Troll, C. (ed.) (1968) Geo-ecology of the mountainous regions of the tropical Americas. *Colloquium Geographicum* Bd. 9, Bonn.

(d)   South-east Asia
Costin, A.B. (1968) Alpine ecosystems of the Australasian region. pp. 55-88 in: *Arctic and Alpine Environments*. Wright, H.E. & Osburn, W.H. (ed.) Bloomington, Indiana.
Steenis, C.G.J. van (1962) The mountain flora of the Malaysian tropics. *Endeavour* 21, pp. 183-193.

INDEX OF SPECIES

SUBJECT INDEX